MASTERING DATA ANALYSIS WITH PYTHON

A Comprehensive Guide to NumPy, Pandas, and Matplotlib

RAJENDER KUMAR

Trademarks

Screenshots

Website References

All website references were current at the date of publication.

For more information, contact: support@JambaAcademy.com.

Published by:
Jamba Academy

Printed in the United States of America

First Printing Edition, 2023

Found Typos & Broken Link

We apologize in advance for any typos or broken link that you may find in this book. We take pride in the quality of our content and strive to provide accurate and useful information to our readers. Please let us know where you found the typos and broken link (if any) so that we can fix them as soon as possible. Again, thank you very much in advance for bringing this to our attention and for your patience.

If you find any typos or broken links in this book, please feel free to email me.

support@JambaAcademy.com

Support

We would love to hear your thoughts and feedback! Could you please take a moment to write a review or share your thoughts on the book? Your feedback helps other readers discover the books and helps authors to improve their work. Thank you for your time and for sharing your thoughts with us!

If there is anything you want to discuss or you have a question about any topic of the book, you can always reach out to me, and I will try to help as much as I can.

support@JambaAcademy.com

To all the readers who have a passion for programming and technology, and who are constantly seeking to learn and grow in their field.

This book is dedicated to you and to your pursuit of knowledge and excellence.

DISCLAIMER

ACKNOWLEDGMENTS

I would like to express my heartfelt gratitude to my colleagues, who provided valuable feedback and contributed to the development of the ideas presented in this book. In particular, I would like to thank Kalpita Dapkekar for her helpful suggestions and support.

I am also grateful to the editorial and production team at Jamba Academy for their efforts in bringing this book to fruition. Their professionalism and expertise were greatly appreciated.

I also want to thank my family and friends for their love and support during the writing process. Their encouragement and understanding meant the world to me.

Finally, I would like to acknowledge the many experts and thought leaders in the field of data science whose works have inspired and informed my own. This book is the culmination of my own experiences and learning, and I am grateful to the wider community for the knowledge and insights that have shaped my thinking.

This book is a product of many people's hard work and dedication, and I am grateful to all of those who played a role in its creation.

HOW TO USE THIS BOOK?

Here are some suggestions for making the most out of the book:

I. **Start with a brief introduction:** To grasp the main goal and organization of the book, start by reading the introduction. You will be better able to comprehend the context and flow of the information presented as a result.

II. **Read the chapters in order**: The chapters are arranged logically and build upon one another to provide readers a thorough comprehension of the subject. To properly understand the principles offered, it is advised to read the chapters in the order they presented.

III. **Put the examples and exercises to use**: To aid readers in understanding and putting the principles covered to use, the book provides examples, exercises, and case studies. Make sure to complete them in the order they are presented in the book.

IV. **Reference the supplementary resources**: To give readers more information and support, the book includes a number of resources, including websites, books, and online courses. Use these tools to help you learn more and stay up to date on what's going on in the field.

V. **Use the information**: Using the knowledge offered in the book in real-world situations is the greatest method to fully comprehend and remember it. In order to get practical experience and consolidate your comprehension, try to use the principles we've covered in your own work or on personal projects.

VI. **Review the chapter summary and questions**: The summary and questions at the end of each chapter are meant to help you review and assess your comprehension of the subject. Before beginning the following chapter, make sure to go over them again.

VII. **Ask for assistance if necessary**: Don't be afraid to ask for assistance if you're having trouble grasping a concept or need further support. Join online forums, go to meetups, or look for a mentor to help you get beyond any challenges you may face.

You may make the most of this book and obtain a thorough understanding of data analysis and its applications in the real world by using these suggestions.

CONVENTIONS USED IN THIS BOOK

When learning a new programming language or tool, it can be overwhelming to understand the syntax and conventions used. In this book, we follow certain conventions to make it easier for the reader to follow along and understand the concepts being presented.

Italics

Throughout the book, we use italics to indicate a command used to install a library or package. For example, when we introduce the Keras library, we will italicize the command used to install it:

!pip install keras

Bold

We use bold text to indicate important terminology or concepts. For example, when introducing the concept of **neural networks**, we would bold this term in the text.

Handwriting Symbol

At times, we may use a handwriting symbol to indicate an important note or suggestion. For example, we may use the following symbol to indicate that a certain code snippet should be saved to a file for later use:

This is an important note or information.

Code Examples

All code examples are given inside a bordered box with coloring based on the Notepad++ Python format. For example:

```
import tensorflow as tf
from tensorflow import keras
from tensorflow.keras import layers
```

```
import numpy as np
import matplotlib.pyplot as plt

# Load the dataset
(x_train, y_train), (x_test, y_test) =
keras.datasets.cifar10.load_data()

# Keep only cat and dog images and their labels
train_mask = np.any(y_train == [3, 5], axis=1)
test_mask = np.any(y_test == [3, 5], axis=1)
x_train, y_train = x_train[train_mask], y_train[train_mask]
x_test, y_test = x_test[test_mask], y_test[test_mask]
```

Output and Explanation

Below each code example, we provide both the output of the code as well as an explanation of what the code is doing. This will help readers understand the concepts being presented and how to apply them in their own code.

Overall, by following these conventions, we hope to make it easier for readers to follow along and learn the concepts presented in this book.

GET CODE EXAMPLES ONLINE

"Mastering Data Analysis with Python: A Comprehensive Guide to NumPy, Pandas, and Matplotlib" is a book written by Rajender Kumar, aimed at individuals looking to enhance their data analysis skills. The book provides a comprehensive guide to the Python libraries NumPy, Pandas, and Matplotlib, which are commonly used for data analysis tasks.

The book is divided into several chapters, each of which covers a different topic in data analysis. The first chapter introduces the NumPy library, which is used for numerical computing tasks such as array manipulation, linear algebra, and Fourier analysis. The subsequent chapters cover the Pandas library, which is used for data manipulation and analysis tasks such as data cleaning, merging, and aggregation.

To make it even more convenient for readers, we are offering all the code discussed in the book as Jupyter notebooks on the link:

https://github.com/JambaAcademy/Mastering_data_Analysis

Jupyter notebooks provide an interactive computing environment that enables users to write and run code, as well as create visualizations and documentation in a single document. This makes it a perfect tool for learning and experimenting with machine learning and deep learning concepts.

The code provided on the Github repository can be downloaded and used freely by readers. The notebooks are organized according to the chapters in the book, making it easier for readers to find the relevant code for each concept.

We believe that this initiative will help readers to gain a better understanding of data cleaning and data analysis concepts by providing them with practical examples that they can run and experiment with.

CONTENTS

MASTERING DATA ANALYSIS WITH PYTHON

ABOUT THE AUTHOR

Rajender Kumar, a data scientist, and IT specialist has always been interested by the use of data to generate insightful conclusions and decision-making. He has a background in both data analysis and computer science, giving him a solid foundation in the technical abilities needed to succeed in these sectors.

But Rajender's passions extend beyond merely the technical facets of their work. Also, he is very interested in the ethical and philosophical implications of artificial intelligence as well as ethical and sustainable technological societal advancement. Rajender's curiosity about the larger effects of technology has inspired him to research subjects like spirituality and mindfulness because he thinks a comprehensive approach to problem-solving is essential in the quickly developing field of data and AI. He likes to meditate and research different spiritual traditions in his free time to achieve inner calm and clarity.

Rajender worked on a variety of initiatives during the course of his career, from creating predictive models for big businesses to creating unique data pipelines for fledgling startups. In each of these projects, he put a strong emphasis on not only providing technical solutions but also closely collaborating with clients to comprehend their needs from a business perspective and support them in achieving their objectives.

Rajender Kumar is a devoted traveler and reader in addition to his professional endeavors. In his own time, he enjoys reading and traveling to new places where he can immerse himself in the local culture.

In both his professional and personal lives, Rajender is continually seeking new challenges and chances to learn and improve. He values the value of lifelong learning and growth, whether it be through remaining current on new advances in his industry or discovering new interests outside of work.

In addition to adding to the continuing discussion about the role of data and technology in our world, he is eager to continue imparting his knowledge and skills to others. He is dedicated to being a thought leader in his industry and having a positive influence on the world around them, whether through writing, speaking, or just having thoughtful conversations with his peers.

OTHER WORK BY THE SAME AUTHOR

- Python Machine Learning: A Beginner's Guide to Scikit-Learn (ISBN: 978-1-960833-01-3)
- The Bhagavad Gita: A Guide to Living with Purpose: A Search for Meaning in Modern Life
- Nugget of Wisdom from the Quran: Life lesson and teaching of the Holy Quran for a peaceful and happy life

WHO THIS BOOK IS FOR?

Anyone with an interest in data analysis and its many practical uses will find much to like in this book. This book is useful for anyone, from those with little prior knowledge of data analysis to more advanced students and practitioners who wish to further their understanding of the subject.

This book serves as an excellent introduction to data analysis for those who are just getting started in the discipline. The straightforward style of the writing makes it ideal for readers who are just getting started with the topic.

In-depth discussions of data visualization, statistical analysis, and exploratory data analysis are provided in this book for readers with prior knowledge in the field. Case studies and real-life examples show how the concepts are used in the actual world.

This book is a great tool for experts in the area to use in order to keep up with the newest research and innovations. New approaches to data analysis are discussed, as are the most up-to-date tools and methods.

This book is perfect for anyone who need to examine data in order to make decisions, such as students, professionals, and researchers in the domains of business, finance, marketing, healthcare, and technology. Because it introduces the fundamental ideas and practices necessary for a career in data analysis or data science, it is also appropriate for anyone considering such a path.

Data analysts, data scientists, business analysts, and data engineers will all find useful information in this book. For anyone tasked with making sense of data, this book is an indispensable resource due to the thorough overview it provides of the data analysis process and the tools and strategies needed to do so.

If you want to learn more about data analysis, its practical applications, and the best ways to use data to make smart decisions and propel your organization forward, this book is for you.

WHAT ARE THE REQUIREMENTS? (PRE-REQUISITES)

To get the most out of the information presented in this book, readers need be aware of a few essential prerequisites.

A primary requirement is a familiarity with elementary statistics and data analysis. Although this book serves as a thorough introduction to these ideas and methods, a familiarity with the subject matter is recommended.

In addition, you should have some programming knowledge, as Python is the major tool for data analysis in this book. The examples and exercises in the book assume that you are already familiar with the fundamentals of Python, including variables, data types, and control flow.

In terms of hardware and software, it is recommended that readers have a Python development environment installed. This includes a text editor or integrated development environment (IDE) for editing and running code, as well as a Python interpreter. NumPy, pandas, and Matplotlib are just a few of the libraries and packages that will be utilized throughout the text. The pip package manager in Python makes it simple to install these packages.

The final step is for readers to get access to some sample data. While the book makes extensive use of real-world datasets in its examples and exercises, readers are also urged to use their own data or locate publically available datasets online to put what they've learned into practice.

Overall, this book is designed to be readable by a wide variety of readers and offers all the knowledge and tools required to build a thorough understanding of the subject matter, even though some prior familiarity with statistics, programming, and data analysis is required.

PREFACE

The subject of data analysis is expanding and changing quickly, and in today's data-driven society, the capacity to successfully analyze and comprehend data is becoming an increasingly valuable skill. Python is a potent and adaptable programming language that is ideal for tasks involving data analysis, and its popularity in the industry has been rising gradually in recent years.

The goal of this book, "Mastering Data Analysis with Python," is to offer a thorough overview of the methods and tools used in data analysis with Python. From the fundamentals of the Python programming language and its built-in data structures to more complex ideas like data visualization, statistical analysis, and data exploration, it covers it all. This book will give you the information and abilities you need to efficiently analyze and interpret data using Python, regardless of your level of data analysis experience.

The book is broken up into a number of chapters, each of which focuses on a different aspect of data analysis. The Python language and its inbuilt data structures, functions, and files are introduced in the first chapter. Data visualization using the well-liked Matplotlib toolkit is covered in the later chapters of the book. Descriptive statistics, including measures of central tendency, distribution, and form, are covered in the seventh chapter. Data analysis in business, including data governance, data quality, business intelligence, and reporting, is covered in the eleventh chapter. Further study materials, including as books, websites, discussion forums, online classes, and data analysis tools, are provided in the appendix A. Appendix B offers advice on how to succeed at data analysis as well as how to land a job as a data analyst.

You'll discover thorough explanations of the ideas and methods discussed in the book, along with illustrations and exercises to help you put what you've learned into practice. A glossary of vocabulary, multiple-choice questions, and a section on best practices for data analysis are also included in the book.

The author of this book has 11 years of experience in the data industry and is passionate about data science, machine learning, analysis, and data integration. Readers will benefit from the author's in-depth expertise and experience in the field of data analysis, which will offer insightful advice.

In conclusion, "Mastering Data Analysis with Python" is the definitive manual for anyone looking to learn and master the skill of Python-based data analysis. This book will give you the information and abilities you need to efficiently analyze and interpret data using Python, regardless of your level of data analysis experience. So, this book is for you if you're ready to delve into the field of data analysis and advance your abilities.

WHY SHOULD YOU READ THIS BOOK?

"Are you curious about how to extract valuable insights from large and complex data sets? Are you looking to improve your data analysis skills using the powerful Python programming language? If so, then Mastering Data Analysis with Python is the book for you.

This comprehensive guide provides a thorough introduction to the various techniques and tools used in data analysis, including data governance, data quality, business intelligence and reporting, and more. Whether you are a beginner or an experienced data analyst, you will find valuable information in this book that will help you take your skills to the next level.

Throughout the book, you will learn how to use Python to perform various data analysis tasks, including data cleaning, visualization, and statistical analysis. You will also learn how to use popular data analysis libraries such as Pandas, Numpy, and Matplotlib.

This book is written in a clear and concise manner, making it easy to understand even for those with no prior experience in data analysis or Python programming. The author's extensive experience in the field is evident in the practical examples and hands-on exercises included throughout the book, which help to solidify your understanding of the concepts covered.

This book is designed to be a practical resource that you can refer back to again and again as you continue to build your data analysis skills. The topics are presented in a logical and easy-to-follow order, and the end-of-chapter summaries and review questions help you to assess your understanding of the material.

In addition, the book includes real-world case studies and examples, as well as tips for success in data analysis. By the end of this book, you will have the knowledge and skills to effectively analyze data and make data-driven decisions in your business or organization.

Rajender Kumar

MASTERING DATA ANALYSIS WITH PYTHON

A COMPREHENSIVE GUIDE TO

NUMPY, PANDAS, AND MATPLOTLIB

01

1 INTRODUCTION TO DATA ANALYSIS WITH PYTHON

Are you ready to dive into the realm of Python data analysis? In this first chapter of our book, 'Mastering Data Analysis with Python,' we will guide you through the intriguing and interesting world of data analysis. You will learn about the data analysis method, different types of data, and numerous data analysis use cases. But that's not all; you'll also learn about the power of Python as a data analysis tool and how to set up your Python environment to begin your data analysis journey. You will be prepared with the required tools to begin your data exploration after learning about key Python data analysis packages such as NumPy, Pandas, and Matplotlib. So, let's get started and follow your curiosity!

1.1 UNDERSTANDING THE BASICS OF DATA ANALYSIS

Data analysis is the process of extracting insights and knowledge from data using various methodologies. It is an important phase in the data science process that assists businesses, organizations, and individuals in making educated decisions based on the data they have. Understanding the data, cleaning and preparing the data, examining and visualizing the data, and lastly generating conclusions from the data are the fundamentals of data analysis.

Understanding the data is the first stage in data analysis. This includes determining the type of data, the sources of the data, and the data format. Structured data, which is arranged in a specific format, such as a database, and unstructured data, which is not organized in a specific format, such as text or photos, are two forms of data. Understanding the data also entails comprehending the environment in which the data was acquired as well as any potential biases that may exist in the data.

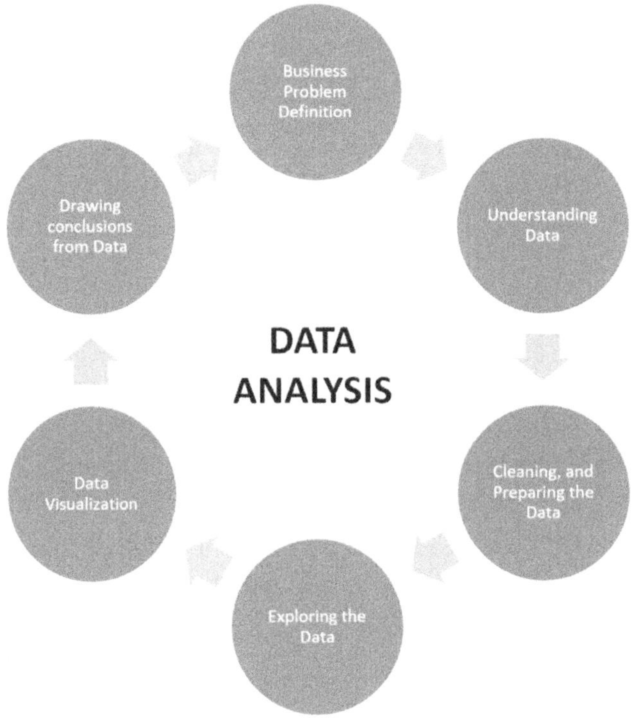

The next stage is to clean and prepare the data for analysis when the data has been understood. Data wrangling is the process of cleaning up errors and inconsistencies from the data, dealing with missing or duplicate data, and converting the data into a format that can be used for analysis. A vital part of the data analysis process is data wrangling, which makes sure the data is correct and trustworthy.

It's time to investigate and visualize the data after it has been prepped and cleansed. Several strategies will be used in this step to comprehend the distribution, trends, and linkages in the data. Insights and patterns in the data that may not be immediately visible can be found by using data exploration and visualization. This step frequently makes use of methods like descriptive statistics, data visualization, and interactive visualizations.

Conclusions from the data are drawn as the final step in data analysis. Making informed decisions and forecasts entails utilizing the understanding and insights received from the data. If new data is gathered or the context of the data changes, conclusions formed from the data may need to be revised. Data analysis is a continual process.

To sum up, data analysis is an important step in the data science process that entails comprehending the data, preparing and cleaning the data, exploring and visualizing the data, and deriving conclusions from the data. It is a never-ending process that calls for attention to detail, a keen mind, and a desire to investigate and experiment with the data. Data analysis has become a valuable talent for people and businesses to make data-based decisions as a result of technological improvements and the growing availability of data.

1.2 TYPES OF DATA

When beginning a data analysis project, one of the most crucial considerations is the type of data. Data can be categorized into three primary categories: structured, unstructured and semi-structured data. If you don't know what kind of data you're dealing with, you can't hope to clean, prepare, and analyze it effectively.

Structured Data

The term "structured data" indicates data that has been meticulously organized in a predetermined system. It's data that can be quickly saved, retrieved, and altered in a computer system. Most structured information consists of numbers and can be simply tallied. Spreadsheets, databases, and comma-separated value (CSV) files are all examples of structured data.

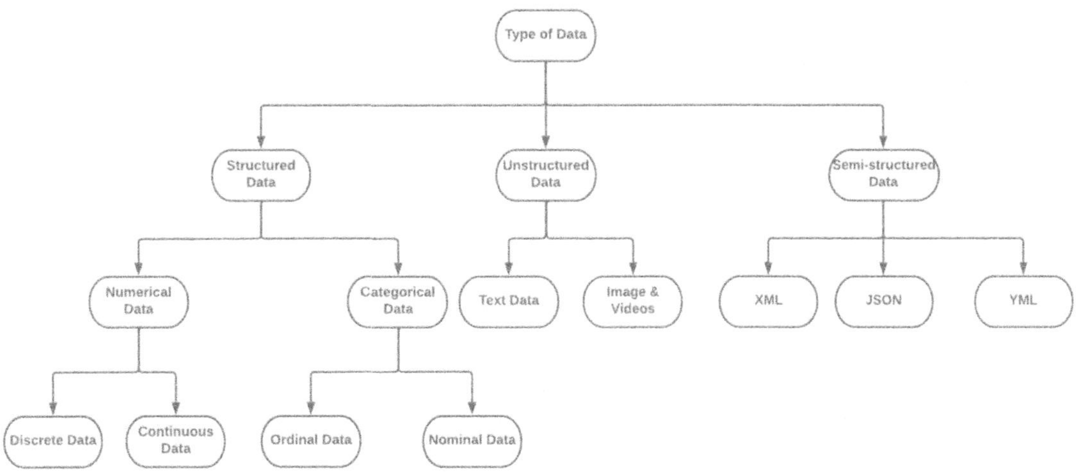

Structured data can be further classified into two types: numerical and categorical.

Numerical data: It is any data that can be represented by numbers. There are two additional categories that can be applied to it: **discrete** and **continuous**. The number of children in a household, for example, is an example of discrete data because it can only take on the values 1, 2, 3, 4, or 5. In contrast, the temperature, which is an example of continuous data, can take on any value within its range.

Categorical Data: Data organized into distinct types, or "categories," is known as "categorical data." Gender, employment, and nationality are all examples of such distinctions. There are two subcategories of categorized information: **ordinal** and **nominal**. Education level is an example of ordinal data because it may be ranked (high school, undergraduate, graduate). Information like gender is an example of nominal data because it cannot be ranked (male, female).

One of the most common forms of structured data is **relational data**, which is data that is organized into tables with rows and columns. Relational data is often stored in a relational database management system (RDBMS) and can be queried using SQL (Structured Query Language). This type of data is commonly used in business and financial applications, such as accounting and inventory management systems.

Transactional data, which is data created as a byproduct of a transaction or interaction, is another type of structured data. Customer contacts with a company can be monitored and trends in customer behavior can be uncovered with the use of transactional data. Purchase records, internet analytics, and call center logs are all examples of transactional data.

Unstructured Data

The term "unstructured data" is used to describe information that does not conform to a predetermined data structure. Unstructured data is harder to quantify and evaluate, but it can be just as useful. Social media, electronic communication, and comments from customers are all examples of unstructured data.

Text data is a prevalent type of unstructured data. Customer comments, social media posts, and other kinds of textual communication can all be mined for useful information using text data. Natural language processing (NLP) methods can be applied to text data in order to extract meaningful insights.

Images are another type of unstructured data. Object recognition, facial recognition, and medical imaging are just few of the many uses for image data. Computer vision techniques allow for the analysis of image data, allowing for the discovery of previously unknown information.

Data in the form of podcasts, movies, and other **audio & video** files is another example of unstructured data. Speech recognition and natural language processing methods can be applied to audio and video data for analysis.

Semi-Structured Data

Semi-structured data is a type of data that has a structure, but not as rigid as structured data. It combines the characteristics of structured and unstructured data, and it is a combination of both. It contains tags or other markers to separate semantic elements and enforce hierarchies of records and fields within the data. XML and JSON are examples of semi-structured data, as they contain a set of rules for organizing data, but also allow for flexibility in the structure.

Web-based data sources, such as web pages and social media feeds, frequently contain semi-structured data. Data from Internet of Things (IoT) devices, like sensor data or log files, may also provide this information. Due to its less rigid nature, semi-structured data can be more difficult to deal with, but it also has the potential to yield greater insights.

The **JSON (JavaScript Object Notation)** format is widely used for storing and exchanging semi-structured data. JSON is widely used in application programming interfaces (APIs) and for transferring data between servers and online applications (Application Programming Interfaces). With Python or R packages, JSON data may be readily converted to structured data, making it more manageable.

Data in **XML** format (Extensible Markup Language) is another example of semi-structured information. Information exchanged over web services frequently takes the form of XML data. In order to make XML data more manageable, it can be modified using XSLT (Extensible Stylesheet Language Transformations).

In conclusion, proper data analysis requires familiarity with the specifics of the data at hand. Because of its convenience, structured data is widely employed in economic and commercial contexts. Although unstructured data is more of a challenge to process, it can be just as informative. The Internet of Things (IoT) and other online data sources provide semi-structured data that combines the two types of data. While cleaning, processing, and analyzing data, it is important to first identify the type of data you are working with.

1.3 SOURCE OF DATA

Data can come from a variety of sources, and understanding the source of the data is crucial for effective data analysis. The source of the data can affect the quality and reliability of the data, and it is important to check the data for errors, inconsistencies, and missing values.

Internal Data Sources

Internal data sources are data that is generated within an organization. This type of data is often stored in databases or spreadsheets and can be used for a variety of purposes, such as accounting and inventory management systems. Examples of internal data sources include:

- Sales data from a company's point-of-sale system
- Customer data from a company's CRM (customer relationship management) system
- Inventory data from a company's inventory management system

One of the advantages of internal data sources is that the data is often of high quality and is well-maintained. The data is also easily accessible and can be used for a variety of purposes. However, internal data sources may not always be complete or may be biased.

External Data Sources

External data sources are data that is generated outside of an organization. This type of data can be found in a variety of sources, such as social media, government websites, and external databases. Examples of external data sources include:

- Social media data, such as tweets or Facebook posts
- Economic data from government websites, such as the Bureau of Labor Statistics
- Demographic data from the United States Census Bureau

One of the advantages of external data sources is that they provide a wide range of information and can be used to gain insights into a variety of areas. However, the quality and reliability of the data from external sources can vary, and it may be more difficult to access and work with.

Crowdsourced Data

Crowdsourced data is data that is collected from a large group of people, usually through an online platform. This type of data can be used for a variety of purposes, such as sentiment analysis and market research. Examples of crowdsourced data include:

- Product reviews on Amazon or Yelp
- Survey data collected through SurveyMonkey or Qualtrics
- Data collected through citizen science projects, such as Zooniverse

One of the advantages of crowdsourced data is that it can provide a large amount of data quickly and at a low cost. However, the quality and reliability of the data can vary, and it may be difficult to verify the accuracy of the data.

Big Data

Big data refers to a large amount of data that is difficult to process and analyze using traditional methods. This type of data is often generated by IoT (Internet of Things) devices and can be found in a variety of sources, such as social media and sensor data. Examples of big data include:

- Sensor data from IoT devices, such as weather stations or traffic cameras
- Social media data, such as tweets or Facebook posts
- Data from web logs, such as server logs or clickstream data

One of the advantages of big data is that it can provide a large amount of information and insights into a variety of areas. However, working with big data can be challenging, as it requires specialized tools and techniques for processing and analyzing the data. Additionally, big data can also be difficult to clean and prepare for analysis, as it may contain a high number of errors, inconsistencies, and missing values.

We should always check the data for errors, inconsistencies, and missing values, regardless of the source of the data. By understanding the source of the data, organizations and individuals can make more informed decisions based on the data they have.

1.4 FORMAT OF DATA

The format of data is another important aspect to consider when embarking on a data analysis project. The format of data refers to the way in which the data is stored, and it can affect how the data is cleaned, prepared and analyzed. Understanding the format of the data will help you determine the appropriate tools and techniques for working with the data.

Tabular Data

Tabular data, also known as relational data, is data that is organized into tables with rows and columns. This type of data is often stored in a relational database management system (RDBMS) and can be queried using SQL (Structured Query Language). Examples of tabular data include data from spreadsheets and CSV files.

Date collected	Plot	Species	Sex	Weight
1/9/78	1	DM	M	40
1/9/78	1	DM	F	36
1/9/78	1	DS	F	135
1/20/78	1	DM	F	39
1/20/78	2	DM	M	43
1/20/78	2	DS	F	144
3/13/78	2	DM	F	51
3/13/78	2	DM	F	44
3/13/78	2	DS	F	146

One of the advantages of tabular data is that it is easy to work with and can be easily converted to other formats, such as a pandas dataframe in Python or a data frame in R. Additionally, tabular data can be easily cleaned, prepared and analyzed using a variety of tools and techniques. However, tabular data can be limited in terms of the types of insights that can be gained from the data.

Text Data

Text data is unstructured data that is often used to extract insights from customer feedback, social media posts, and other forms of written communication. Text data can be found in a variety of formats, such as plain text files, PDFs, and HTML files. Examples of text data include customer reviews, news articles, and social media posts.

One of the advantages of text data is that it can provide a wealth of information and insights into a variety of areas. Text data can be analyzed using natural language processing (NLP) techniques, such as sentiment analysis and topic modeling, to extract insights from the data. However, text data can be difficult to work with, as it is unstructured, and it can be time-consuming to clean and prepare the data for analysis.

Image Data

Image data is unstructured data that is often used in applications such as object recognition, facial recognition, and medical imaging. Image data can be found in a variety of formats, such as JPEG, PNG, and TIFF. Examples of image data include photos, scanned documents, and medical images.

One of the advantages of image data is that it can provide a wealth of information and insights into a variety of areas. Image data can be analyzed using computer vision techniques, such as object detection and image segmentation, to extract insights from the data. However, image data can be large in size and can be difficult to work with, as it requires specialized tools and techniques for processing and analyzing the data.

Audio and Video Data

Audio and video data is unstructured data that can be found in forms of podcast, videos, and other audio files. Audio and video data can be found in a variety of formats, such as MP3, WAV, and MPEG. Examples of audio and video data include podcast, videos, and speech recordings.

One of the advantages of audio and video data is that it can provide a wealth of information and insights into a variety of areas. Audio and video data can be analyzed using speech recognition and natural language processing techniques to extract insights from the data. However, audio and video data can be large in size and can be difficult to work with, as it requires specialized tools and techniques for processing and analyzing the data.

In conclusion, understanding the format of the data is crucial for effective data analysis. Data can come in a variety of formats, including tabular data, text data, image data and audio and video data.

1.5 BENEFITS OF DATA ANALYSIS

Data analysis is the process of examining, cleaning, transforming, and modeling data to extract useful information and insights. It is a critical component in today's business and research world. With the vast amount of data that is being generated every day, it is essential to have the tools and techniques to make sense of it. In this section, we will explore the benefits of data analysis and why it is important.

- **Improved Decision Making:** One of the most important benefits of data analysis is that it enables better decision making. By analyzing data, businesses and organizations can gain a deeper understanding of their customers, products, and operations. This information can then be used to make informed decisions that lead to increased revenue, reduced costs, and improved performance.
- **Identification of Trends and Patterns:** Data analysis helps in identifying trends and patterns in the data that might not be immediately obvious. This information can be used to identify new opportunities or to develop new products and services. For example, by analyzing sales data, a business might identify a trend of increased demand for a particular product. This information can be used to increase production or to develop a new product line.
- **Predictive Analysis:** Data analysis also enables predictive analysis, which allows businesses and organizations to make predictions about future events. This can be used for forecasting, risk management, and resource planning. Predictive models can be used to identify potential problems and opportunities before they occur, which can help organizations to be more proactive in their decision making.

- **Cost Reduction:** Data analysis can also be used to reduce costs. By identifying areas of inefficiency or waste, organizations can make changes that lead to cost savings. For example, by analyzing production data, a manufacturing company might identify a machine that is causing delays and increasing costs. By replacing or repairing that machine, the company can reduce costs and improve efficiency.
- **Improved Customer Experience:** Data analysis can be used to improve the customer experience. By analyzing customer data, businesses can gain a deeper understanding of their customers' needs, preferences, and behavior. This information can be used to develop more effective marketing campaigns, improve products and services, and provide better customer support.
- **Better allocation of resources:** Data analysis can also be used to improve the allocation of resources. By analyzing data on resource utilization, an organization can identify areas where resources are being wasted and make changes that lead to more efficient resource allocation.
- **Compliance:** Data analysis can also be used to comply with regulations and laws. By analyzing data, organizations can identify potential compliance issues and take steps to address them before they become a problem. This can help to avoid costly fines and legal action.
- **Better understanding of the market:** Data analysis can also be used to gain a better understanding of the market. By analyzing data on market trends, competitors, and customer behavior, organizations can identify opportunities and threats, and develop strategies to stay ahead of the competition.
- **Improved Product and Service Development:** Data analysis can be used to improve product and service development. By analyzing data on customer needs, preferences, and behavior, organizations can develop products and services that are better aligned with customer needs. This can lead to increased sales and customer satisfaction.
- **Streamlining Business Processes:** Data analysis can be used to streamline business processes. By analyzing data on process performance, organizations can identify bottlenecks and inefficiencies. By making changes to the processes, organizations can improve efficiency and reduce costs.
- **Personalized Marketing:** Data analysis can be used to develop personalized marketing strategies. By analyzing data on customer behavior and preferences,

organizations can develop targeted marketing campaigns that are more likely to be successful.

- **Data-Driven Innovation:** Data analysis can also be used to drive innovation. By analyzing data on customer needs, preferences, and behavior, organizations can identify new opportunities for product and service development.
- **Better Resource Management:** Data analysis can be used to manage resources more effectively. By analyzing data on resource utilization, organizations can identify areas where resources are being wasted and make changes to improve resource management.
- **Improved Fraud Detection:** Data analysis can also be used to detect fraud. By analyzing data on financial transactions, organizations can identify patterns of fraudulent activity and take steps to prevent it.

In conclusion, data analysis is a powerful tool that can be used to improve decision making, identify trends and patterns, make predictions, reduce costs, improve customer experience, better allocate resources and comply with regulations. With the vast amount of data that is being generated every day, it is essential for businesses and organizations to have the tools and techniques to make sense of it, and that's what data analysis is all about.

1.6 DATA ANALYSIS USE CASES

Data analysis is a powerful tool that can be used to extract insights and knowledge from data, and it has a wide range of use cases across various industries. In this section, we will explore several common use cases of data analysis.

Business Intelligence

Business intelligence (BI) is the process of using data to make better business decisions. Data analysis plays a key role in BI by providing organizations with insights into their operations and performance. This can include analyzing sales data to identify trends and patterns, analyzing customer data to identify areas for improvement, and analyzing financial data to identify areas for cost savings.

One of the most common use cases for BI is customer segmentation. Customer segmentation is the process of dividing customers into groups based on common characteristics. This can be used to identify the most profitable customers, target marketing efforts, and improve customer service.

Marketing Analytics

Marketing analytics is the process of using data to understand and improve marketing efforts. Data analysis plays a key role in marketing analytics by providing insights into customer behavior, marketing campaigns, and the effectiveness of marketing channels. This can include analyzing customer data to identify patterns in customer behavior, analyzing marketing campaign data to identify areas for improvement, and analyzing social media data to identify areas for engagement.

One of the most common use cases for marketing analytics is A/B testing. A/B testing is the process of comparing two versions of a marketing campaign to determine which one is more effective. This can be used to test different headlines, images, and call-to-action buttons to determine which elements of a campaign are most effective.

Fraud Detection

Fraud detection is the process of identifying fraudulent activity in financial transactions. Data analysis plays a key role in fraud detection by identifying patterns and anomalies in financial transactions. This can include analyzing transaction data to identify patterns of fraudulent activity, analyzing customer data to identify high-risk customers, and analyzing network data to identify connections between fraudulent transactions.

One of the most common use cases for fraud detection is credit card fraud detection. Credit card companies use data analysis to identify patterns of fraudulent activity, such as a large number of small transactions or transactions that occur outside of a customer's normal spending patterns.

Predictive Maintenance

Predictive maintenance is the process of using data to predict when equipment or machinery will need maintenance. Data analysis plays a key role in predictive maintenance by providing insights into the health and performance of equipment. This can include analyzing sensor data to identify patterns in equipment performance, analyzing maintenance data to identify areas for improvement, and analyzing failure data to identify the root cause of equipment failure.

One of the most common use cases for predictive maintenance is the use of sensor data to predict when equipment will fail. This can include analyzing vibration data to identify patterns of equipment wear, analyzing temperature data to identify patterns of overheating, and analyzing pressure data to identify patterns of equipment failure.

Retail

In the field of retail, data analysis is used to optimize inventory management, track sales, and identify customer buying trends. Retailers use data analysis to optimize their supply chains, minimize waste, and improve their bottom line.

Healthcare

Another industry where data analysis plays a crucial role is healthcare. Medical professionals use data analysis to identify patterns and trends in patient data, such as medical history, symptoms, and test results. This helps them to make more accurate diagnoses and develop more effective treatment plans. In addition, data analysis is used to track the spread of diseases, monitor the effectiveness of treatments, and identify potential health risks.

Data analysis is a powerful tool that can be used to extract insights and knowledge from data, and it has a wide range of use cases across various industries. Business intelligence, marketing analytics, fraud detection, and predictive maintenance are just a few examples of the many ways that data analysis can be used to improve decision making and drive business success. Data analysis is a key tool for organizations looking to make data-driven decisions and stay competitive in today's data-driven world.

1.7 SUMMARY

- Data analysis is the process of examining, cleaning, transforming, and modeling data to extract useful information and insights.
- Data analysis is an iterative process that involves multiple steps, including data acquisition, data cleaning, data exploration, data visualization, and model building.
- Data can come from a variety of sources, including databases, files, and the web.
- Data can be in a variety of formats, including structured, semi-structured, and unstructured.

- Data exploration and visualization is a critical step in the data analysis process, as it allows for the discovery of patterns, trends, and relationships in the data.
- Data analysis can be used to support decision making, identify new opportunities, and generate insights for various fields such as business, healthcare, social science, and more.
- Python is a popular programming language for data analysis, and it has several libraries and frameworks such as NumPy, Pandas, Matplotlib, and Scikit-learn that make it easier to perform data analysis tasks.
- The goal of data analysis is to extract insights and information from data to support decision-making and problem-solving.
- Data analysis can be applied to different types of data, including numerical, categorical and time series.
- Understanding the basics of data analysis is crucial to be able to work with data and extract insights from it.
- Data analysis can be performed using various methods and techniques, including descriptive statistics, inferential statistics, and machine learning.
- Descriptive statistics provide a summary of the data's main characteristics, such as the mean, median, and standard deviation.
- Inferential statistics allow for drawing conclusions about a population based on a sample of data.
- Machine learning is a subset of artificial intelligence that allows for the building of models that can learn from data and make predictions.
- Data visualization is a key component of data analysis, as it allows for the communication of insights and patterns in a clear and easily understood way.
- Data visualization tools such as Matplotlib, Seaborn, and Plotly can be used to create various types of plots and charts.
- Data analysis is an ongoing process, as new data is constantly being acquired and added to the dataset. It requires constant monitoring, validation and maintenance.
- Data analysis is an essential skill for data scientists, data analysts, and business analysts, as well as professionals in many other fields.

1.8 TEST YOUR KNOWLEDGE

I. What is the main purpose of data analysis?

a) To extract insights and knowledge from data
b) To organize data into tables
c) To improve the design of a website
d) To create charts and graphs

II. **What is the most common form of structured data?**
a) JSON
b) XML
c) CSV
d) Excel

III. **What is the main advantage of internal data sources?**
a) They provide a wide range of information
b) They are easily accessible
c) They are of high quality and well-maintained
d) They are unbiased

IV. **What is the main advantage of external data sources?**
a) They provide a wide range of information
b) They are easily accessible
c) They are of high quality and well-maintained
d) They are unbiased

V. **What type of data is used for sentiment analysis?**
a) Tabular data
b) Text data
c) Image data
d) Audio and video data

VI. **What is the main advantage of crowdsourced data?**
a) They provide a large amount of data quickly and at a low cost
b) They are of high quality and well-maintained
c) They provide a wide range of information
d) They are easily accessible

VII. **What is the main advantage of big data?**
a) They provide a large amount of information and insights into a variety of areas
b) They are easily accessible
c) They are of high quality and well-maintained
d) They are unbiased

VIII. **What type of data is used for object recognition?**

a) Tabular data
b) Text data
c) Image data
d) Audio and video data

IX. **What is the main advantage of image data?**

a) They provide a wealth of information and insights into a variety of areas
b) They are easily accessible
c) They are of high quality and well-maintained
d) They are unbiased

X. **What is the main use case of predictive maintenance?**

a) To predict when equipment or machinery will need maintenance
b) To identify patterns of fraudulent activity
c) To make better business decisions
d) To improve marketing efforts

1.9 ANSWERS

I. Answer: a) To extract insights and knowledge from data.
II. Answer: d) Excel
III. Answer: c) They are of high quality and well-maintained
IV. Answer: a) They provide a wide range of information
V. Answer: b) Text data
VI. Answer: a) They provide a large amount of data quickly and at a low cost
VII. Answer: a) They provide a large amount of information and insights into a variety of areas
VIII. Answer: c) Image data
IX. Answer: a) They provide a wealth of information and insights into a variety of areas
X. Answer: a) To predict when equipment or machinery will need maintenance

02

2 GETTING STARTED WITH PYTHON

This chapter serves as the foundation for the rest of the book as we delve into mastering data analysis with Python. Before we dive into the complexities of data analysis, we must first ensure that we have a solid understanding of the basics of Python and how to set up the environment to run our code.

In this chapter, we will guide you through the process of setting up your Python environment, whether it be on your local machine or in the cloud. We will also cover the fundamental concepts of Python such as variables, data types, operators, and control flow structures. By the end of this chapter, you will have the necessary tools and knowledge to begin your journey towards mastering data analysis with Python.

Whether you are new to programming or have experience with other programming languages, this chapter will provide a comprehensive introduction to the basics of Python and set you up for success as you continue through the rest of the book. So, let's get started and dive into the world of Python and data analysis!

2.1 INSTALLING PYTHON

Installing Python is the first step in setting up your environment for data analysis and programming. Python is a widely used, high-level programming language that is versatile and easy to learn. It is an interpreted language, which means that the code is executed line by line, rather than being compiled. This makes it an ideal language for data analysis, as it allows for rapid prototyping and iteration.

There are two main versions of Python that are currently in use: Python 2 and Python 3. Python 2 was released in 2000, and Python 3 was released in 2008. While both versions are still in use,

Python 3 is the recommended version to use, as it has several improvements over Python 2. Therefore, in this section, we will focus on installing Python 3.

Installing Python on Windows:

- Download the Python installer from the official Python website: https://www.python.org/downloads/windows/

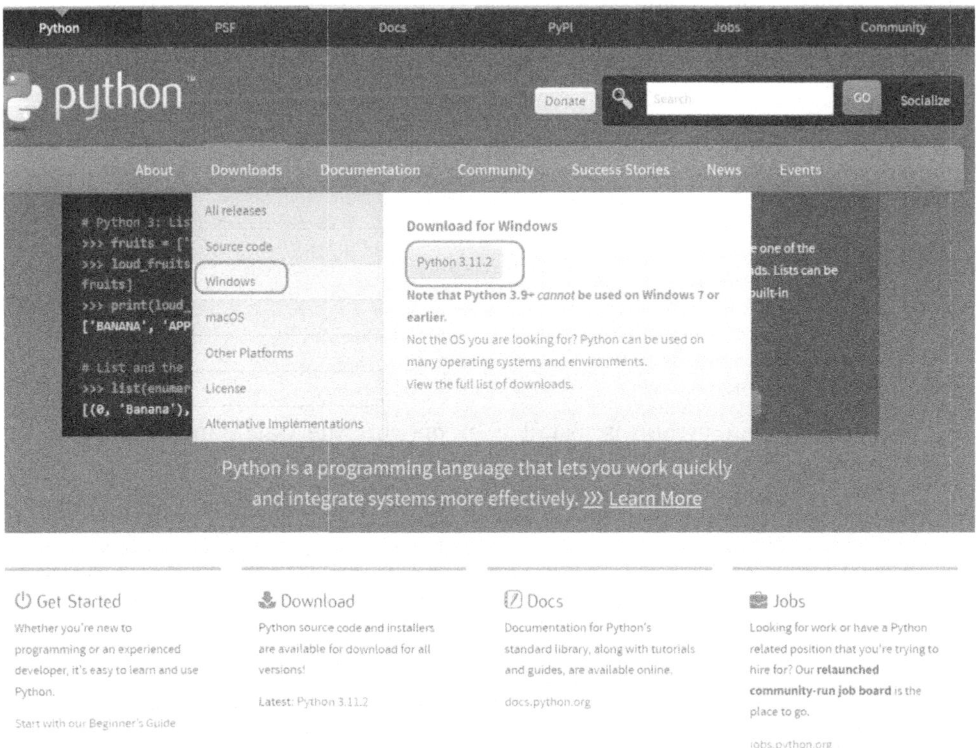

- Run the installer and follow the prompts to install Python
- Make sure to check the box to add Python to your system path

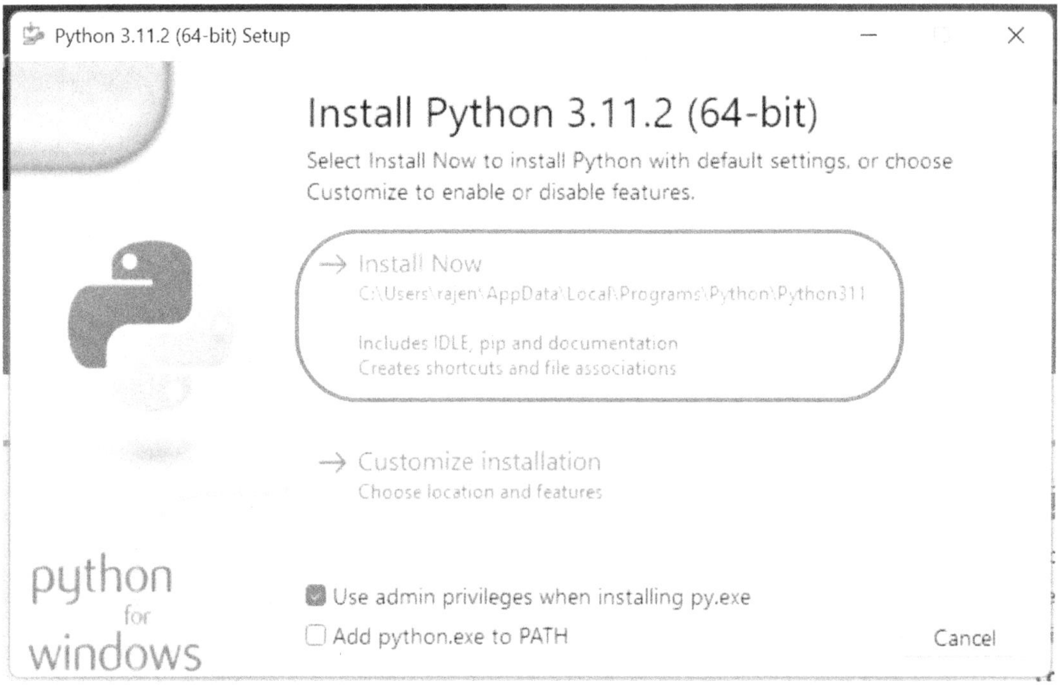

- Verify that Python is installed by opening the Command Prompt and typing "python"

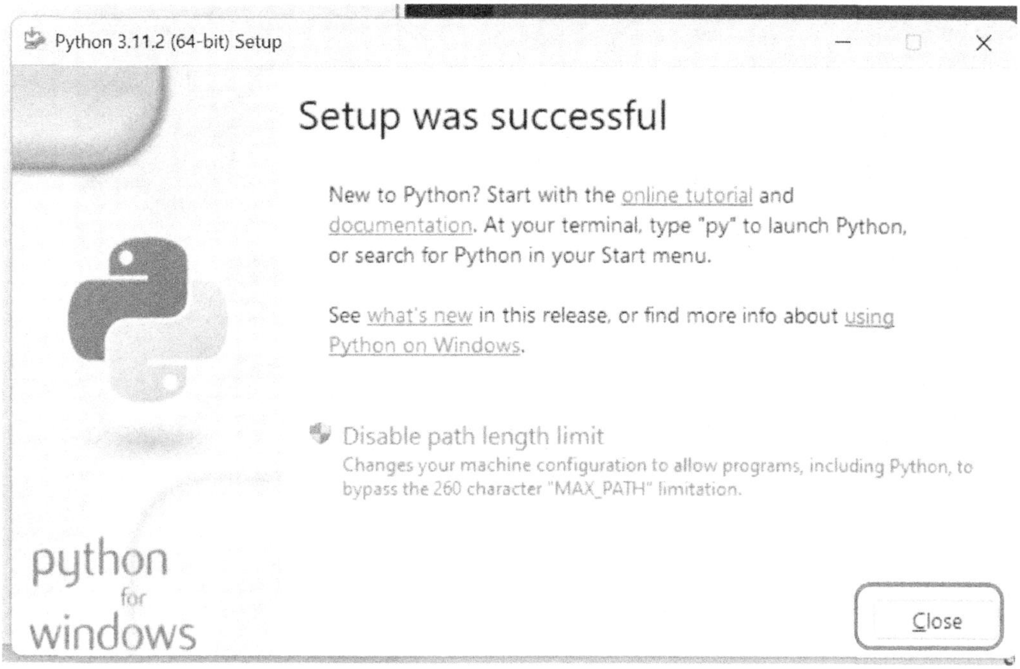

Installing Python on Mac:

- Download the Python installer from the official Python website: https://www.python.org/downloads/mac-osx/

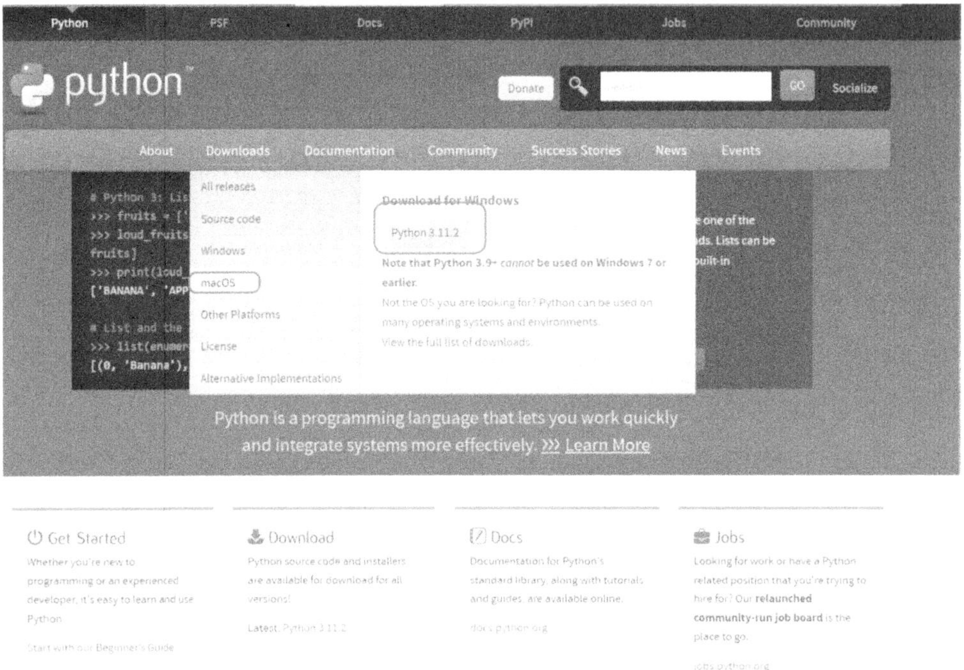

- Run the installer and follow the prompts to install Python
- Verify that Python is installed by opening the Terminal and typing "python3"

Installing Python on Linux:

- Use the package manager that comes with your Linux distribution to install Python
- For Ubuntu and Debian: open the Terminal and type "sudo apt-get install python3"
- For Fedora and Red Hat: open the Terminal and type "sudo yum install python3"
- Verify that Python is installed by opening the Terminal and typing "python3"

Installing packages with pip:

- pip is the package installer for Python and it is included with Python 3
- To install a package, use the command "*pip install package_name*"
- To update a package to the latest version, use the command "*pip install --upgrade package_name*"
- For example, to install pandas library, use the command "*pip install pandas*"

Please follow the above steps to install python, and if you face any issues, you can check the python documentation for more information.

2.2 SETTING UP JUPYTER NOTEBOOK

Jupyter Notebook is an open-source web application that allows you to create and share documents that contain live code, equations, visualizations, and narrative text. It is a popular tool for data analysis, as it allows you to easily combine code, data, and visualizations in a single document. In this section, we will go through the process of setting up Jupyter Notebook and show you how to use it for data analysis.

Installing Jupyter Notebook

To install Jupyter Notebook, you will first need to have Python installed on your computer. If you do not already have Python installed, you can refer to the previous section on installing Python for instructions. Once Python is installed, you can use the package manager pip to install Jupyter Notebook. Open the command prompt or terminal and type "*pip install jupyter*" and press Enter. This will install the latest version of Jupyter Notebook on your computer.

Starting Jupyter Notebook

Once Jupyter Notebook is installed, you can start it by opening the command prompt or terminal and typing "*jupyter notebook*" and press Enter. This will launch the Jupyter Notebook web application in your default web browser. The web application will show the files and folders in the current directory, and you can navigate to the directory where you want to create a new notebook.

Creating a new notebook

To create a new notebook, click on the "New" button and select "Python 3" from the drop-down menu. This will create a new notebook with a single empty cell. You can add new cells by clicking on the "Insert" button or by using the keyboard shortcut "B" (for "below") or "A" (for "above").

Working with Cells

Cells are the basic building blocks of a Jupyter Notebook. They can contain text or code, and you can use them to organize your notebook into sections. There are two types of cells: Markdown cells and Code cells. **Markdown cells** are used to display text, while **Code cells** are used to run code.

Markdown cells allow you to write formatted text, such as headings, lists, and links. To change a cell to a Markdown cell, click on the drop-down menu in the toolbar and select "Markdown". You can then enter your text and use markdown syntax to format it.

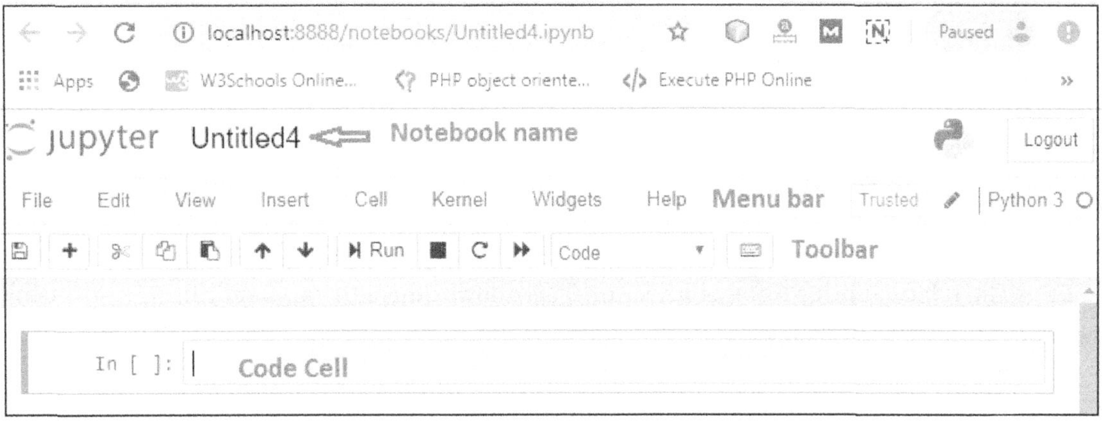

Code cells, on the other hand, allow you to write and run code. To change a cell to a Code cell, click on the drop-down menu in the toolbar and select "Code". You can then enter your code and run it by clicking the "Run" button or using the keyboard shortcut "Shift + Enter". The output of the code will be displayed below the cell.

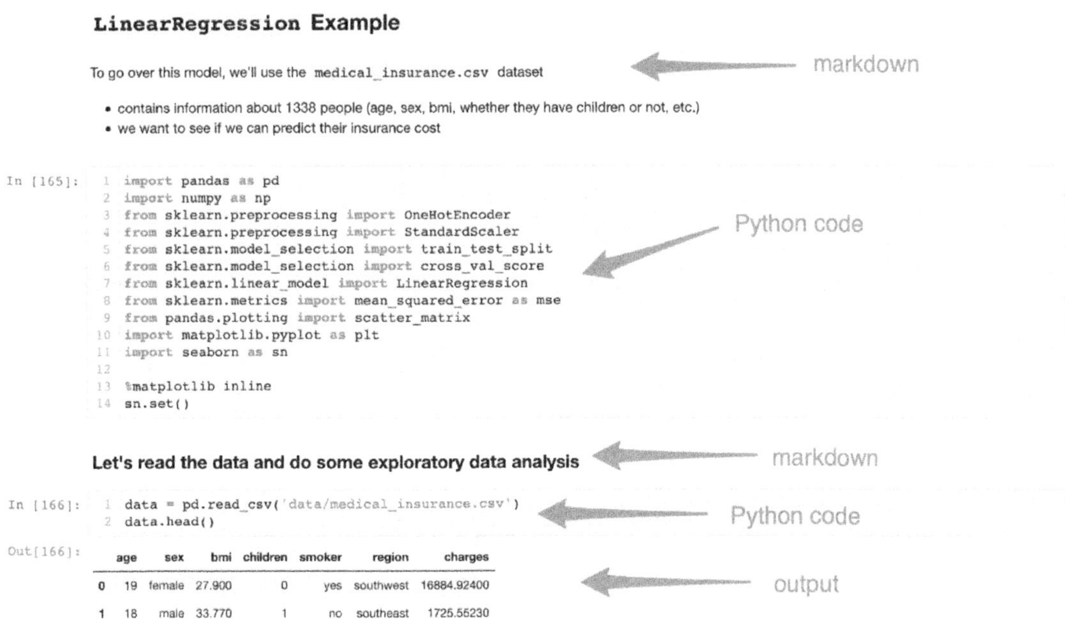

You can also add comments to your code by using the "#" symbol. These comments will not be executed when you run the code, and they are useful for explaining what the code is doing.

Exporting and Sharing Notebooks

Jupyter Notebook allows you to export your notebook to various formats, such as HTML, PDF, and Python script. This allows you to share your notebook with others, even if they do not have Jupyter Notebook installed. To export your notebook, go to the "File" menu and select "Download As". You can then select the format you want to export the notebook in.

Additionally, Jupyter Notebook also allows you to share your notebook online using services such as GitHub or Jupyter Notebook Viewer. This allows others to view and run your notebook without having to download it.

Jupyter Notebook is a powerful tool for data analysis, as it allows you to combine code, data, and visualizations in a single document. It is easy to set up and use, and it provides a wide range of features that make it useful for data analysis tasks. By following the steps outlined in this section, you should now be able to set up Jupyter Notebook on your computer and start using it for data analysis.

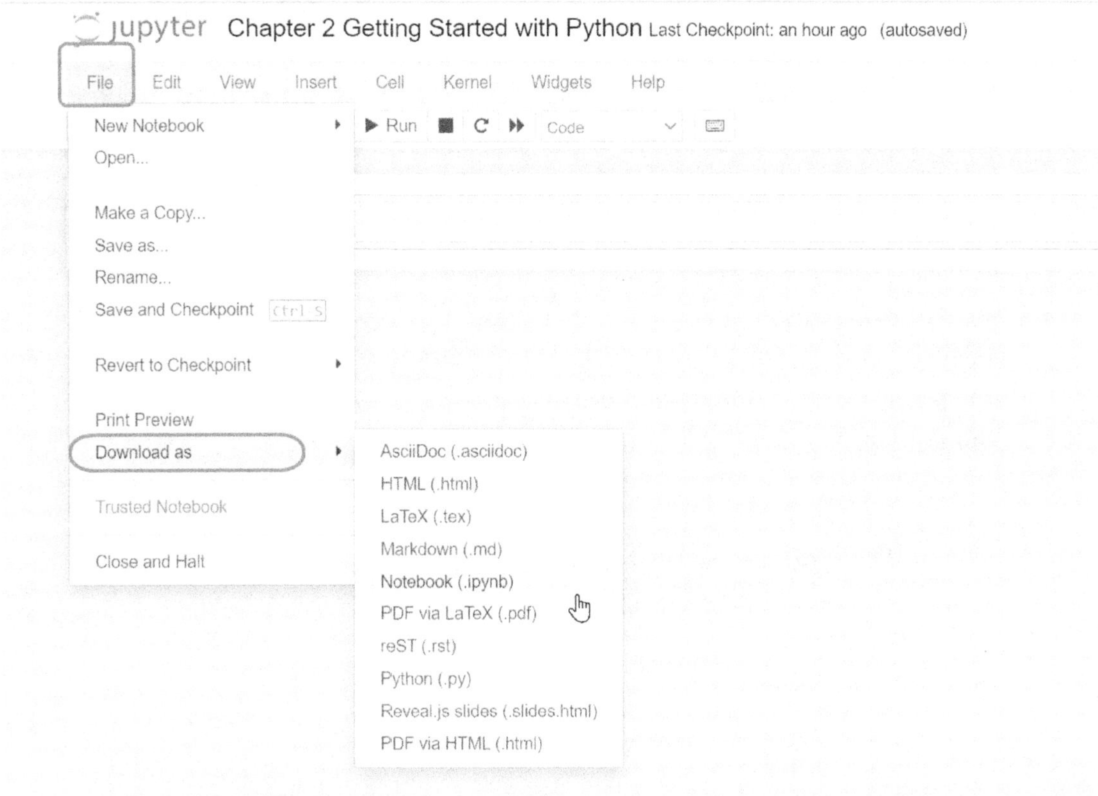

2.3 MAGIC COMMANDS IN JUPYTER

Magic commands are special commands that are not part of the Python language and are used to perform special actions in Jupyter Notebook. These commands are prefixed with "%" or "%%" and are used to perform various tasks, such as running code in a different language, displaying the execution time of a cell, and controlling the behavior of Jupyter Notebook.

%magic commands

When working with Jupyter notebooks, there are many useful features and functionalities that can streamline your workflow and make your life easier. One of the most powerful tools at your disposal is the %magic command, a set of built-in commands that allow you to perform a wide range of tasks directly from your Jupyter notebook.

At its core, the %magic command is a way to interact with the Python interpreter and the Jupyter notebook environment. By using these commands, you can quickly and easily perform a variety of tasks, such as changing the behavior of the notebook, importing data, executing shell commands, and more.

Here are some examples of %magic commands that you might find useful in your data analysis work:

- **%matplotlib inline**: This command enables inline plotting of figures in your Jupyter notebook, allowing you to create and view visualizations directly in your code cells. For example, to create a scatter plot of two variables named 'x' and 'y', you can use the following code:

```
In [1]:
%matplotlib inline
import matplotlib.pyplot as plt

x = [1, 2, 3, 4, 5]
y = [2, 4, 6, 8, 10]

plt.scatter(x, y)
plt.show()
```

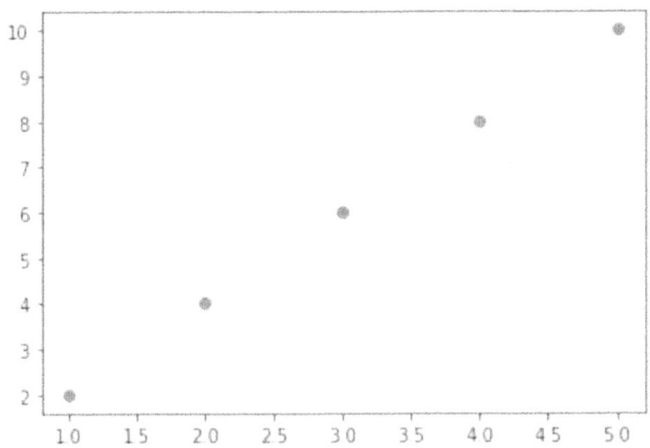

- **%load filename**: This command loads the contents of a file into a code cell, allowing you to edit and run the code directly from the notebook. For example, to load code from a file named 'my_code.py', you can use the following code:

```
In [ ]:  %load my_code.py
```

- **%run filename**: This command runs the specified Python script in a separate process, allowing you to execute complex or long-running code without tying up your Jupyter notebook. For example, to run a script named 'my_script.py', you can use the following code:

```
In [ ]:  %run my_script.py
```

- **%timeit statement**: This command measures the execution time of a Python statement or expression, allowing you to optimize your code and improve performance. For example, to time the execution of a for loop that appends elements to a list, you can use the following code:

```
In [3]:  %timeit
         my_list = []
         for i in range(1000):
             my_list.append(i)

```

- %reset: This magic command allows you to reset the namespace by removing all names defined by the user. For example, to reset the namespace, you can use the following code:

```
In [6]:  %reset


Once deleted, variables cannot be recovered. Proceed (y/[n])? y
```

- **%who and %whos**: These commands display information about the variables in your current namespace, helping you keep track of your data and ensure that your code is running correctly.

```
In [5]:  %whos

Variable   Type      Data/Info
------------------------------
i          int       999
my_list    list      n=1000
plt        module    <module 'matplotlib.pyplo<...>\\matplotlib\\pyplot.py'>
x          list      n=5
y          list      n=5
```

These are just a few examples of the many %magic commands available in Jupyter notebooks. To see a full list of available commands, simply type **%lsmagic** in a code cell and run the cell.

```
%lsmagic

Available line magics:
%alias  %alias_magic  %autoawait  %autocall  %automagic  %autosave  %bookmark  %cd  %clear  %cls  %colors  %conda  %config  %co
nnect_info  %copy  %ddir  %debug  %dhist  %dirs  %doctest_mode  %echo  %ed  %edit  %env  %gui  %hist  %history  %killbgscripts
%ldir  %less  %load  %load_ext  %loadpy  %logoff  %logon  %logstart  %logstate  %logstop  %ls  %lsmagic  %macro  %magic  %matpl
otlib  %mkdir  %more  %notebook  %page  %pastebin  %pdb  %pdef  %pdoc  %pfile  %pinfo  %pinfo2  %pip  %popd  %pprint  %precisio
n  %prun  %psearch  %psource  %pushd  %pwd  %pycat  %pylab  %qtconsole  %quickref  %recall  %rehashx  %reload_ext  %ren  %rep
%rerun  %reset  %reset_selective  %rmdir  %run  %save  %sc  %set_env  %store  %sx  %system  %tb  %time  %timeit  %unalias  %unl
oad_ext  %who  %who_ls  %whos  %xdel  %xmode

Available cell magics:
%%!  %%HTML  %%SVG  %%bash  %%capture  %%cmd  %%debug  %%file  %%html  %%javascript  %%js  %%latex  %%markdown  %%perl  %%prun
%%pypy  %%python  %%python2  %%python3  %%ruby  %%script  %%sh  %%svg  %%sx  %%system  %%time  %%timeit  %%writefile

Automagic is ON, % prefix IS NOT needed for line magics.
```

One of the most powerful aspects of the %magic command is its ability to customize and extend the Jupyter notebook environment. By creating your own %magic commands, you can automate repetitive tasks, build custom workflows, and streamline your data analysis workflow.

For example, let's say that you frequently work with data stored in CSV files. You could create a custom %magic command that automatically loads a CSV file into a Pandas DataFrame, allowing you to quickly and easily analyze the data using the power of Python and the Jupyter notebook environment.

To create a custom %magic command, you can use the IPython Magics API, which provides a simple interface for defining and registering new commands. With a little bit of Python code, you can create a powerful and flexible tool that can save you time and make your data analysis work more efficient.

%%magic commands

%%magic commands are cell-level commands that are prefixed with a "%%" symbol. These commands are used to perform actions that affect the entire cell. Some examples of %%magic commands include:

- **%%timeit**: The %%time command allows users to time how long it takes for a particular cell to execute. This can be useful for optimizing code and identifying potential bottlenecks in a program. For example, let's say we want to time how long it takes to create a list of random numbers:

```
%%time
import random
numbers = [random.randint(1, 100) for _ in range(1000000)]

```

```
CPU times: total: 594 ms
Wall time: 853 ms
```

- %%html: The %%html command allows users to display HTML content in a Jupyter notebook cell. This can be useful for creating rich content, such as tables or charts. For example, let's say we want to display a simple HTML table:

```
%%html
<table>
  <tr>
    <th>Name</th>
    <th>Age</th>
  </tr>
  <tr>
    <td>John</td>
    <td>30</td>
  </tr>
  <tr>
    <td>Jane</td>
    <td>25</td>
  </tr>
</table>

```

Name	Age
John	30
Jane	25

The output will display the HTML table in the Jupyter notebook cell as above.

- **%%capture**: The **%%capture** magic command in Jupyter notebook is used to capture the standard output and error streams of a cell. This can be particularly useful when dealing with code that produces a lot of output or when running long processes that may have error messages.
- Here's an example of how to use the **%%capture** magic command:

```
%%capture captured_output

print("This is some example output")
for i in range(5):
    print(i)

print("Done with output")

```

```
print(captured_output.stdout)
```

```
This is some example output
0
1
2
3
4
Done with output
```

In this example, we use **%%capture** to capture the output of the cell, which includes the text "This is some example output", the numbers 0-4, and "Done with output". We assign this captured output to the variable **captured_output**.
We can then access the captured output by calling **captured_output.stdout** or **captured_output.stderr**, depending on which stream we want to access. For example:

- **%%writefile**: The %%writefile command allows users to write the contents of a cell to a file. This can be useful for saving code or data for later use. For example, let's say we want to save a Python script called "my_script.py" that contains a function called "my_function":

```
%%writefile my_script.py
def my_function():
    print("Hello, world!")

```

```
Writing my_script.py
```

- **%%latex**: The **%%latex** magic command in Jupyter Notebook is used to render LaTeX code within a cell. LaTeX is a high-quality typesetting system used to create technical and scientific documents. With this magic command, you can input LaTeX code directly into a cell and the output will be displayed as formatted text.

  To use **%%latex**, simply begin a cell with **%%latex** and then input your LaTeX code within the cell. Once you run the cell, the output will be displayed as formatted text.

  Here's an example:

```
%%latex

\begin{equation}
    E = mc^2
\end{equation}

This is the famous equation derived by Albert Einstein.
```

$$E = mc^2$$

This is the famous equation derived by Albert Einstein.

As you can see, the LaTeX code within the cell is rendered as formatted text. This is a useful tool for those working with technical and scientific documents who want to input equations or other mathematical symbols directly into a cell.

Using Magic Commands

To use magic commands, you simply need to prefix the command with "%" or "%%" and then run the cell. For example, to measure the execution time of a single line of code, you would use the %timeit magic command.

```
%timeit sum(range(100))
```

Similarly, to measure the execution time of an entire cell, you would use the %%timeit magic command.

```
%%timeit
sum = 0
for i in range(100):
    sum += i
```

It's also possible to pass some optional arguments to the magic commands, for example, to repeat the execution of the command to get a more accurate measure of the execution time, you can use -r (or --repeat) argument in the timeit magic command:

```
%timeit -r 5 sum(range(100))
```

Magic commands are an essential feature of Jupyter Notebook that allows you to perform various actions that are not possible with Python alone. By using these commands, you can easily measure the execution time of your code, run scripts, load files, and much more. The %magic commands are used to perform actions that affect the current line or cell and the %%magic commands are used to perform actions that affect the entire cell. Familiarizing yourself with the magic commands available in Jupyter Notebook will help you to work more efficiently and effectively.

2.4 INSTALLING REQUIRED LIBRARIES

Installing required libraries is an essential step in setting up your environment for data analysis and programming. Libraries are collections of modules that provide additional functionality to your Python environment. They can be used to perform tasks such as data manipulation, visualization, and machine learning. In this section, we will go through the process of installing libraries and show you how to use them in your data analysis projects.

Using pip to install libraries

Pip is the package installer for Python, and it is included with Python 3. It allows you to easily install, update, and remove libraries from your Python environment. To install a library, you can use the command:

"*pip install library_name*".

For example, to install the popular data analysis library pandas, you can use the command "*pip install pandas*". This will install the latest version of pandas, which can be used for data manipulation and cleaning.

You can also use pip to update a library to the latest version. To update a library, you can use the command:

"*pip install --upgrade library_name*".

Installing multiple libraries at once

It is common to use multiple libraries in a data analysis project, and installing them one by one can be time-consuming. To install multiple libraries at once, you can use a requirements file. A requirements file is a plain text file that contains a list of libraries and their version numbers. You can create a requirements file by running the command below in your command prompt or terminal:

"*pip freeze > requirements.txt*"

This command will create a file called "requirements.txt" in your current directory, which will contain the list of libraries and their version numbers that are installed in your environment. You can then use the below command to install all the libraries listed in the file:

"*pip install -r requirements.txt*"

Installing libraries from source

In some cases, you may need to install a library from source, rather than using pip. This may be necessary if the library is not available on the Python Package Index (PyPI) or if you need a specific version of the library. To install a library from source, you will need to download the source code and then use the command "*python setup.py install*" to install it.

Using libraries in your code

Once a library is installed, you can use it in your code by importing it. You can import a library using the "import" keyword, followed by the name of the library.

For example, to use the pandas library in your code, you would use the following line of code:

```
import pandas as pd
```

This line of code imports the pandas library and assigns it the alias "pd", which you can use to refer to the library in your code.

Installing required libraries is an essential step in setting up your environment for data analysis and programming. Pip is the most common tool to install libraries, it's easy to use, and it can install, update, and remove libraries from your Python environment. You can also use a requirements file to install multiple libraries at once or install libraries from source. Once the libraries are installed, you can use them in your code by importing them using the "import" keyword.

Keeping your library versions up to date and using the latest versions of the libraries can provide you with new functionalities and fix bugs. Also, when working with different projects it's a good practice to create a virtual environment for each project to avoid version conflicts and to have a clean environment for each project.

2.5 BASICS OF PYTHON LANGUAGE

Python is a high-level, interpreted programming language that is widely used in scientific computing, data analysis, artificial intelligence, and other fields. It is designed to be easy to read and write, with a simple and consistent syntax that makes it easy for new programmers to learn.

Python is a versatile language, with a wide range of libraries and modules that are available for various tasks. It is also an open-source language, which means that it is free to use and distribute.

In this section, we will cover the basics of the Python language, including its data types, variables, and operators.

Indentation in Python

Indentation in Python refers to the use of whitespace at the beginning of a line to indicate the level of nesting in the code. It is used to define the scope and structure of the code, and is a crucial aspect of the Python programming language.

In Python, indentation is used to indicate the beginning and end of a block of code, such as a loop or a function. The level of indentation determines the level of nesting in the code, with each level of indentation representing a deeper level of nesting. For example, a block of code indented four spaces would be considered to be one level deeper than a block of code indented two spaces.

Indentation is also used to separate statements within a block of code. In Python, each statement must be on its own line, and must be indented at the same level as the surrounding code. This helps to make the code more readable and easier to understand.

It's important to note that in Python, indentation is mandatory, and the use of incorrect indentation can lead to IndentationError. This is why it's important to be consistent and use the same number of spaces for each level of indentation throughout the code.

In addition, many text editors and IDEs have built-in features to automatically indent code, which can help to ensure consistency and reduce the risk of errors.

Comment in Python

Comments are an important aspect of any programming language, as they allow developers to leave notes for themselves and others about the functionality of the code. In Python, comments are represented by a pound sign (#) and can be placed at the beginning of any line or after a statement.

Single-line comments are used to provide a brief description or explanation of the code that follows it. They are placed at the beginning of a line and can be used to explain the purpose of a statement or function. For example:

```
# This is a single-line comment
x = 5 # This is also a single-line comment
```

Multi-line comments, also known as docstrings, are used to provide more detailed explanations of code and are typically used for documenting functions, classes, and modules. They are enclosed in triple quotes (''' or """) and can span multiple lines. For example:

```
def add_numbers(x, y):
    """
    This function takes two numbers as input and returns their sum
    """
    return x + y
```

It is a best practice to include comments in your code to make it more readable and understandable for yourself and others. Comments can also be used to disable a piece of code temporarily without having to delete it.

In python, comments can be used to provide meta-information in form of annotations, which are ignored by the interpreter but are used by various tools like IDEs, linters, and documentation generators.

In addition to these, the PEP 8 style guide for Python recommends that comments should be complete sentences, start with a capital letter and end with a period, and that blank lines should be used to separate them from the code.

Overall, comments are an essential tool for making your code more readable and understandable. They should be used strategically and judiciously to provide helpful information without overwhelming the reader with unnecessary details.

Data Types

In Python, data type refers to the classification or categorization of data items. In other words, data types determine what operations can be performed on a variable and how it can be stored in memory. Python has several built-in data types, including:

1. **Numbers**: The number data type in Python is used to store numeric values. It includes integers, floats, and complex numbers.

Examples:

```
a = 10          # integer
b = 3.14        # float
c = 2 + 3j      # complex number
```

2. **Strings**: The string data type is used to represent a sequence of characters. In Python, strings are enclosed in either single or double quotes.

Examples:

```
name = "John Doe"          # string using double quotes
address = '123 Main St'  # string using single quotes
```

3. **Booleans**: The boolean data type in Python is used to represent True or False values.

Examples:

```
is_raining = True        # boolean with True value
is_sunny = False         # boolean with False value
```

4. **Lists**: A list is an ordered collection of elements, and each element can be of a different data type.

Examples:

```
my_list = [1, "hello", 3.14, True] # a list containing different data types
```

5. **Tuples**: A tuple is similar to a list, but it is immutable (cannot be modified).

Examples:

```
my_tuple = (1, 2, 3) # a tuple of integers
```

6. **Sets**: A set is an unordered collection of unique elements.

Examples:

```
my_set = {1, 2, 3} # a set of integers
```

7. **Dictionaries**: A dictionary is an unordered collection of key-value pairs.

Examples:

```
my_dict = {'name': 'John', 'age': 30, 'address': '123 Main St'}
# a dictionary with string keys and different data types as values
```

These are some of the commonly used data types in Python. It is important to understand the data types in Python to effectively manipulate and process data.

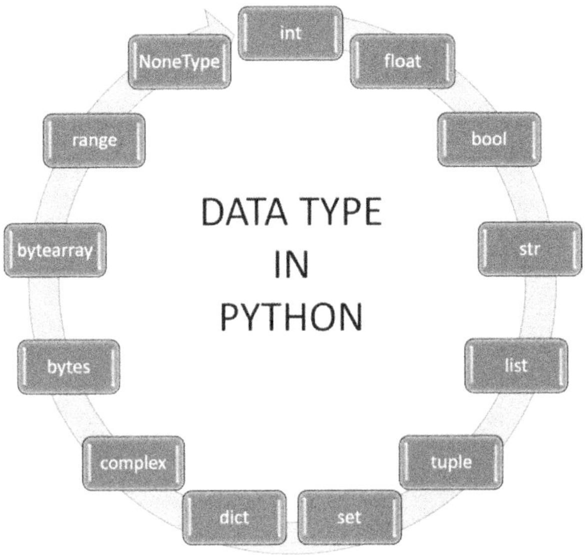

Variables

In Python, a variable is a named memory location that is used to store a value. It is a way to refer to the value stored in memory by giving it a name. Variables are created when a value is assigned to them using the assignment operator (=). Once a variable is created, it can be used in expressions and statements, and its value can be updated or changed.

There are a few rules to follow when naming variables in Python:

- Variable names cannot start with a number
- Variable names can only contain letters, numbers, and underscores
- Variable names cannot be the same as Python keywords (e.g. if, else, while, etc.)

There are a few conventions to follow when naming variables in Python:

- Variable names should be lowercase
- Words in a variable name should be separated by underscores (e.g. my_variable)

In Python, the type of a variable is determined by the value that is assigned to it. Python is a dynamically typed language, which means that the type of a variable can change at runtime. For example, a variable can be assigned an integer value and later be reassigned a string value.

Operator in Python

Operators in Python are special symbols that perform specific operations on one or more operands. An operand is a variable or a value on which the operator performs the operation.

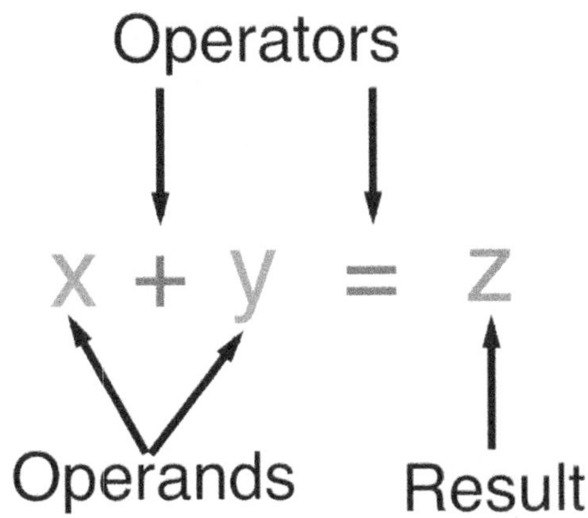

There are several types of operators in Python, including:

- **Arithmetic operators:** These operators perform basic mathematical operations such as addition (+), subtraction (-), multiplication (*), division (/), modulus (%), and exponentiation (**).
- **Comparison operators:** These operators compare two values and return a Boolean value (True or False) depending on the comparison. They include equal to (==), not equal to (!=), greater than (>), less than (<), greater than or equal to (>=), and less than or equal to (<=).
- **Logical operators:** These operators perform logical operations such as and (and), or (or), and not (not). They are used to combine multiple conditions in an if-else statement.

- **Bitwise operators:** These operators perform bit-level operations on integers. They include bitwise and (&), bitwise or (|), bitwise xor (^), bitwise complement (~), and left shift (<<) and right shift (>>) operators.
- **Assignment operators:** These operators are used to assign values to variables. They include the assignment operator (=), the addition assignment operator (+=), the subtraction assignment operator (-=), and so on.
- **Identity operators:** These operators are used to compare the memory location of two variables. They include is and is not.
- **Membership operators**: These operators test whether a value is found in a sequence (such as a string, list, or tuple). They include **in** and **not in**.

Priority of operators in Python follows the standard mathematical order of operations (PEMDAS). Parentheses can be used to change the order of operations, and the use of parentheses is highly recommended to make the code more readable.

2.6 CONTROL FLOW

Control flow in Python refers to the order in which the code is executed. It allows for conditional execution of code, enabling the program to make decisions and take different actions based on certain conditions. The three main control flow structures in Python are:

If-Elif-Else Statements

The if statement in Python is used to make decisions in your code. It allows you to check if a certain condition is true or false, and then execute different code depending on the outcome.

The basic syntax for an if statement is as follows:

```
if condition:
    # code to execute if condition is true
else:
    # code to execute if condition is false
```

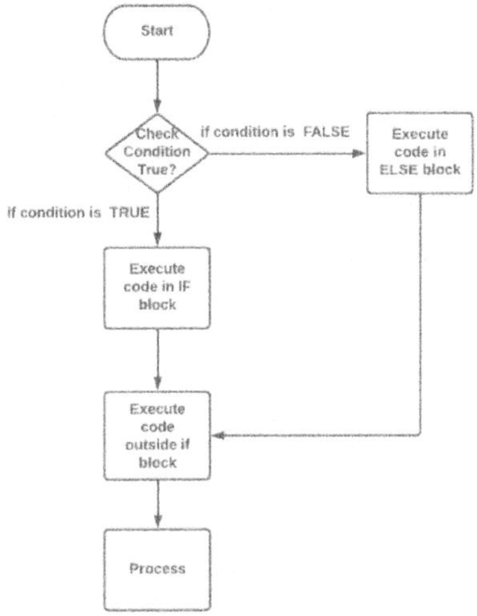

The condition can be any expression that evaluates to either True or False. For example, you can check if a variable is equal to a certain value, or if an element is present in a list.

```
x = 5
if x > 0:
    print("x is positive")
else:
    print("x is non-positive")
```

You can also use elif (short for "else if") to include additional conditions to check.

```
x = 5
if x > 0:
    print("x is positive")
elif x < 0:
    print("x is negative")
else:
    print("x is zero")
```

Code inside the if block will only be executed if the condition is true, and similarly for the else and elif blocks. Also, once a condition is found to be true, the rest of the conditions are not checked.

In addition to making simple decisions, the if statement can also be used in conjunction with loops to perform more complex operations on data. For example, you can use an if statement inside a for loop to filter out certain elements of a list, or use it with a while loop to exit the loop when a certain condition is met.

For Loops

A for loop in Python is a control flow statement that allows you to iterate over a sequence of items, such as a list, tuple, or string. The basic syntax of a for loop is as follows:

```
for variable in sequence:
    # code to be executed for each item in the sequence
```

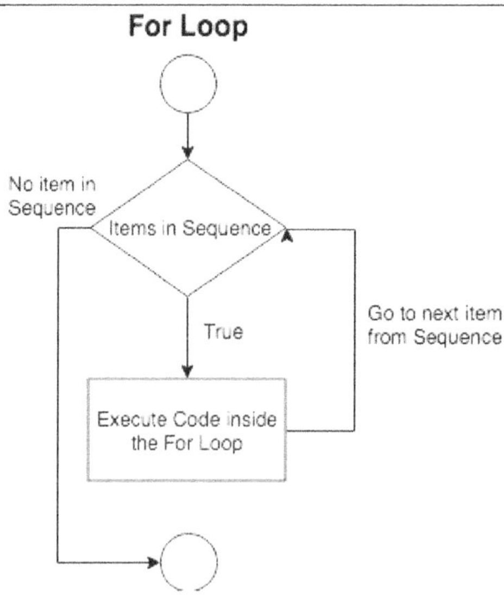

The variable is a placeholder that holds the current item in the sequence, and the code block indented under the for statement is executed for each item in the sequence. For example, the following for loop iterates over a list of integers and prints each one:

```
numbers = [1, 2, 3, 4, 5]
for num in numbers:
    print(num)
```

This would output:

```
1
2
3
4
5
```

It's also possible to use the range() function to generate a sequence of numbers to iterate over. For example, the following for loop prints the numbers from 0 to 9:

```
for i in range(10):
    print(i)
```

This would output:

```
0
1
2
3
4
5
6
7
8
9
```

It's also possible to use the enumerate() function to iterate over a list and also get the index of each item. For example, the following for loop prints the index and value of each item in a list:

```
fruits = ['apple', 'banana', 'orange']
for i, fruit in enumerate(fruits):
    print(i, fruit)
```

This would output:

```
0 apple
1 banana
2 orange
```

In addition, the **break** and **continue** statements can be used within a for loop to control the flow of the loop. The break statement will exit the loop when it is encountered, while the continue statement will skip to the next iteration of the loop.

While Loops

A while loop in Python allows for the execution of a block of code repeatedly as long as a given condition is true. The basic syntax for a while loop is as follows:

```
while condition:
    # code to be executed
```

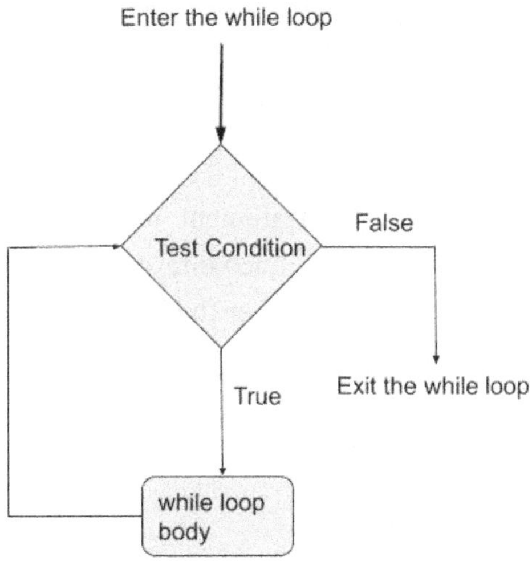

The condition is evaluated before the code block is executed, and if the condition is true, the code block is executed. Once the code block is executed, the condition is evaluated again, and the process repeats until the condition is no longer true.

For example, the following code uses a while loop to print the numbers 1 through 5:

```
i = 1
while i <= 5:
    print(i)
    i += 1
```

This while loop starts with the variable i set to 1. The condition i <= 5 is evaluated, and since it is true, the code block is executed, which prints the value of i to the console. Then the variable i is incremented by 1 (i += 1), and the process repeats until the variable i is no longer less than or equal to 5.

It's important to be careful when using while loops, as if the condition is never false, it can lead to an infinite loop that will crash your program. To avoid this, it's best practice to include a mechanism for ending the loop, such as incrementing a counter variable, or setting a flag variable.

Also, you should be careful with the condition in the while loop, it should not be infinite condition, or it will cause infinite loop, which will crash the program.

Break

The "break" statement is a control flow statement in Python that is used to exit a loop prematurely. When the break statement is encountered in a loop, the loop is immediately terminated and the program control resumes after the loop. The break statement can be used in both "for" loops and "while" loops.

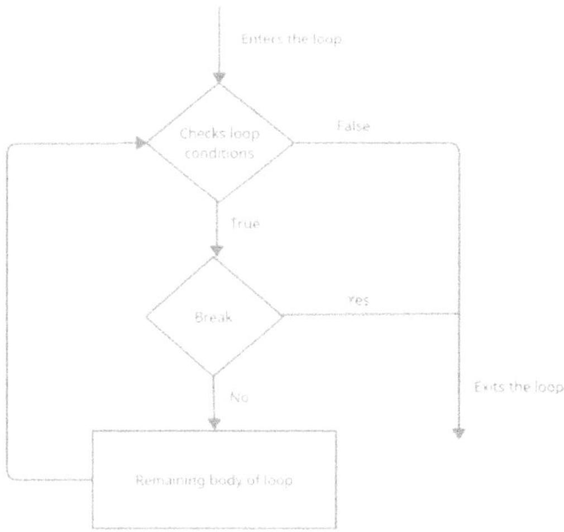

Here is an example of using the break statement in a "for" loop:

```
for i in range(10):
    if i == 5:
        break
    print(i)
```

In this example, the loop will iterate over the range of 0 to 9. However, when the value of "i" reaches 5, the "if" statement will evaluate to True, and the break statement will be executed, causing the loop to exit. The output of this code will be the numbers 0 through 4, as the loop exits before it reaches the value of 5.

The break statement can also be used in a "while" loop:

```
i = 0
while True:
    i += 1
    if i == 5:
        break
    print(i)
```

In this example, the "while" loop will run indefinitely because the condition "True" will always evaluate to True. However, the "if" statement will again check if the value of "i" is equal to 5, and if it is, the break statement will be executed, causing the loop to exit. The output of this code will again be the numbers 0 through 4.

The break statement can also be used in nested loops, where it will exit only the innermost loop it is in.

Break statement should be used with caution and only when necessary, as it can make the code more complex and harder to understand. Additionally, it should be avoided in situations where a "continue" statement would suffice.

Continue

In Python, the continue statement is used to skip over a particular iteration in a loop. This statement is typically used in conjunction with a conditional statement, such as an if statement, in order to control the flow of the loop. When the continue statement is encountered, the current iteration of the loop is terminated and the next iteration begins.

The continue statement is particularly useful when working with large datasets or when performing complex computations, as it can help to reduce the amount of unnecessary processing that needs to be done. For example, if you are iterating over a large list of data and only need to process a subset of the data based on some condition, you can use the continue statement to skip over any data that does not meet the condition.

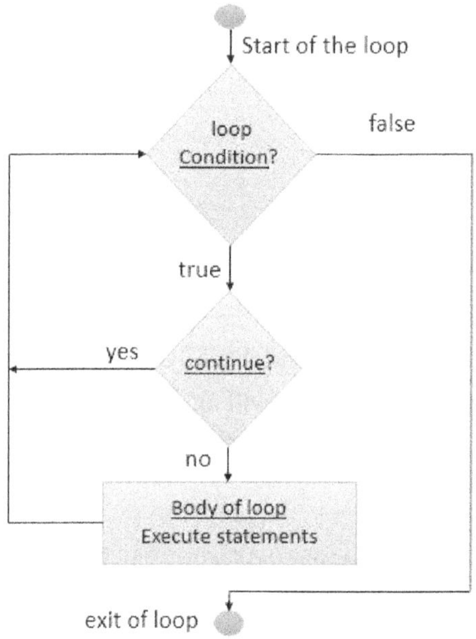

In addition to its use in loops, the continue statement can also be used in other control flow structures, such as try-except blocks, to control the flow of execution in more complex programs. For example, if you have a try-except block that is used to handle exceptions, you can use the continue statement to continue processing even if an exception is encountered.

When used correctly, the continue statement can help to make your code more efficient and easier to read. However, it is important to use it with caution and to thoroughly test your code to ensure that it is functioning as expected.

Pass

In Python, the pass statement is used as a placeholder for code that is yet to be implemented. It serves as a no-operation statement, that is, it does not perform any action or computation when executed. It is often used as a placeholder in control flow statements such as if, for, and while loops, as well as in class and function definitions.

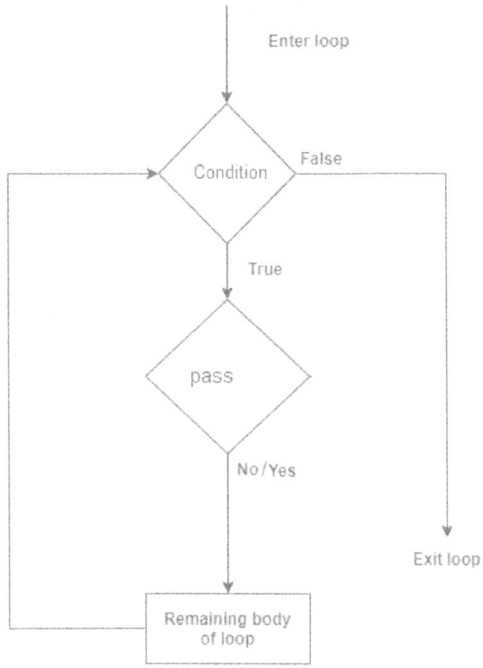

For example, when working on a larger codebase, a developer may want to define a function but not implement its functionality yet. In this case, the developer can use the pass statement to indicate that the function is intended to be used in the future and to prevent the interpreter from raising a syntax error.

Another use case for the pass statement is when creating an empty class or an empty control flow statement. Without the pass statement, the interpreter would raise an error for an empty block of code. Using the pass statement allows the developer to indicate that this block of code is intended to be empty and that it should be ignored.

Indentation plays a crucial role in defining the scope of control flow statements. In python, a block of code is defined by the indentation level, and it is considered a good practice to use four spaces for indentation.

Try to avoid using more than two levels of indentation, and use meaningful variable names. Also, it is a good practice to add comments to your code, explaining what each block of code does, and to use docstrings to document your functions. This will help others to understand your code and also make it easier to maintain in the future.

Control flow in Python is an essential tool for writing efficient and effective code. With a good understanding of the basic control flow structures and best practices, you will be well equipped to write clean, readable, and maintainable code.

2.7 INTRODUCTION TO THE PYTHON DATA ANALYSIS LIBRARIES (NUMPY, PANDAS, MATPLOTLIB)

Python is a popular language for data analysis, and there are several libraries available that make it easy to work with data in Python. In this section, we will introduce three of the most popular data analysis libraries in Python: NumPy, Pandas, and Matplotlib. Although we are introducing these libraries here, we will discuss each of these in great detail in future chapters.

NumPy

NumPy, short for **Numerical Python**, is a library that provides support for large multi-dimensional arrays and matrices of numerical data, as well as a large collection of mathematical functions to operate on these arrays. It is the fundamental package for scientific computing with Python.

One of the main features of NumPy is its N-dimensional array object, or ndarray, which allows you to efficiently store and manipulate large arrays of homogeneous numerical data (i.e., data of the same type, such as integers or floating-point values).

NumPy also provides functions for performing mathematical operations on arrays, such as linear algebra, Fourier transforms, and random number generation.

Pandas

Pandas is a library built on top of NumPy that provides fast and flexible data structures for data analysis. It has two main data structures: **Series** and **DataFrame**. Series is a one-dimensional labeled array capable of holding any data type, while DataFrame is a 2-dimensional labeled data structure with columns of potentially different types.

Pandas provide a wide range of data manipulation functions, such as merging, reshaping, and grouping data. It also has powerful data import and export capabilities, making it easy to work with data in a variety of formats, including CSV, Excel, and SQL.

Matplotlib

Matplotlib is a plotting library for creating static, animated, and interactive visualizations in Python. It provides an object-oriented API for embedding plots into applications using general-purpose GUI toolkits like Tkinter, wxPython, Qt, or GTK.

Matplotlib supports a wide range of plot types, including line plots, scatter plots, bar plots, and histograms, and it can be used to create both 2D and 3D plots. It also has extensive customization options, allowing you to control every aspect of a plot, such as its colors, labels, and legends.

In addition to the features mentioned above, these libraries also have many other capabilities that can be used to solve specific data analysis problems. For example, NumPy has built-in functions for working with polynomials and Fourier transforms, and Pandas has functions for working with time series data. Matplotlib also has a variety of specialized plots, such as heatmaps and 3D plots.

Another great advantage of these libraries is that they are **open-source** and **actively maintained**, which means that they are constantly updated with new features and bug fixes. There is also a large community of users and developers who contribute to these libraries, providing a wealth of resources such as tutorials, documentation, and sample code.

In summary, NumPy, Pandas, and Matplotlib are essential libraries for data analysis in Python. They provide powerful and efficient tools for working with data and offer a wide range of capabilities to solve specific data analysis problems. These libraries are open-source and actively maintained, making them reliable and easy to use for data analysis tasks.

2.8 SUMMARY

- Python is a popular programming language for data analysis and scientific computing.
- To get started with Python, it is necessary to first install it on your computer.
- There are different ways to install Python, including using the Anaconda distribution, which comes with many popular data analysis libraries pre-installed.
- Jupyter Notebook is a popular tool for writing and running Python code, and it is often used in data analysis projects.

- To use Jupyter Notebook, it must first be installed, and then it can be launched from the command line.
- Jupyter Notebook allows for interactive coding and visualization, and it supports many languages including Python, R, and Julia.
- Magic commands are special commands in Jupyter Notebook that are not part of the Python language, but provide additional functionality.
- Magic commands start with a % symbol, and they can be used to perform tasks such as timing code, running shell commands, and more.
- To use magic commands in Jupyter Notebook, it is necessary to have a basic understanding of the command line interface.
- It is important to have a good understanding of the basics of Python, Jupyter Notebook and Magic commands before starting any data analysis project.
- Understanding how to install Python, Jupyter and how to use magic command is a fundamental step to work with data analysis projects in python.
- Magic commands are special commands in Jupyter Notebook that are not part of the Python language, but provide additional functionality.
- Magic commands start with a % symbol and they can be used to perform tasks such as timing code, running shell commands, and more.
- Magic commands are divided into two types: line magics, which are applied to a single line, and cell magics, which are applied to a whole cell.
- Some common line magics include %time, %timeit, and %run, which can be used to time the execution of code, profile the memory usage of code, and run external scripts respectively.
- Some common cell magics include %%time, %%timeit, and %%run, which work similarly to the line magics but apply to the entire cell.
- Magic commands can also be used to run shell commands directly from Jupyter Notebook using %sh or %%sh magics.
- Magic commands can also be used to load external code using %load_ext magic command.
- Magic commands can be very useful for debugging, profiling and running shell command in Jupyter Notebook.
- Understanding how to use magic commands can greatly improve the productivity and efficiency while working with Jupyter Notebook.

2.9 TEST YOUR KNOWLEDGE

Here are ten multiple choice questions to test your understanding:

I. **What is the main feature of the NumPy library?**
   a) Support for large multi-dimensional arrays and matrices of numerical data
   b) Fast and flexible data structures for data analysis
   c) Plotting library for creating static, animated, and interactive visualizations
   d) Random number generation

II. **What is the main data structure of the Pandas library?**
   a) N-dimensional array object
   b) Series and DataFrame
   c) Plotting library for creating static, animated, and interactive visualizations
   d) Random number generation

III. **What is the main purpose of the Matplotlib library?**
   a) Support for large multi-dimensional arrays and matrices of numerical data
   b) Fast and flexible data structures for data analysis
   c) Plotting library for creating static, animated, and interactive visualizations
   d) Random number generation

IV. **What type of data is best suited for the NumPy library?**
   a) Unstructured data
   b) Homogeneous numerical data
   c) Text data
   d) Time series data

V. **Which library can be used to perform data manipulation and cleaning?**
   a) NumPy
   b) Pandas
   c) Matplotlib
   d) All of the above

VI. **Which library allows you to embed plots into applications?**
   a) NumPy
   b) Pandas
   c) Matplotlib
   d) All of the above

VII. **What is the command to install a library using pip?**

a) pip install library_name
b) pip update library_name
c) pip remove library_name
d) pip import library_name

VIII. What type of file is used to install multiple libraries at once?

a) Requirements file
b) Libraries file
c) Data file
d) Script file

IX. What is the command to create a requirements file?

a) pip freeze > requirements.txt
b) pip install > requirements.txt
c) pip list > requirements.txt
d) pip create > requirements.txt

X. What is the command to install a library from source?

a) pip install library_name
b) python setup.py install
c) pip update library_name
d) pip remove library_name

2.10 ANSWERS

I. Answer: a) Support for large multi-dimensional arrays and matrices of numerical data
II. Answer: b) Series and DataFrame
III. Answer: c) Plotting library for creating static, animated, and interactive visualizations
IV. Answer: b) Homogeneous numerical data
V. Answer: b) Pandas
VI. Answer: c) Matplotlib
VII. Answer: a) pip install library_name
VIII. Answer: a) Requirements file
IX. Answer: a) pip freeze > requirements.txt
X. Answer: b) python setup.py install

03

3 BUILT-IN DATA STRUCTURES, FUNCTIONS, AND FILES

The chapter "Built-in Data Structures, Functions, and Files" is an introduction to some of the key concepts and tools in the Python programming language. This chapter will cover the basics of Python data structures, including lists, tuples, sets, and dictionaries. We will also explore built-in Python functions and how to work with files in Python. Understanding these concepts is essential for anyone looking to use Python for data analysis or other related tasks. This chapter will provide a solid foundation for further learning and exploration in the field of Python programming.

3.1 BUILT-IN DATA STRUCTURES

Built-in data structures in Python include lists, tuples, sets, and dictionaries. These data structures can be used to store and organize various types of data in an efficient and convenient manner.

Lists

Python's built-in data structures include lists, which are used to store an ordered collection of items. Lists are similar to arrays in other programming languages, but they are more flexible and powerful.

Creating a List

A list is created by placing a comma-separated sequence of items inside square brackets. For example, the following code creates a list of integers:

```
numbers = [1, 2, 3, 4, 5]
print(numbers)
```

```
numbers = [1, 2, 3, 4, 5]
print(numbers)
```

```
[1, 2, 3, 4, 5]
```

Accessing List Items

You can access individual items in a list by their index. Python uses zero-based indexing, so the first item in the list has an index of 0, the second item has an index of 1, and so on.

```
print(numbers[0]) #prints 1
print(numbers[1]) #prints 2
```

```
print(numbers[0]) #prints 1
print(numbers[1]) #prints 2
```

```
1
2
```

Modifying List Items

You can modify the value of a list item by assigning a new value to its index.

```
numbers[0] = 10
print(numbers)
```

```
numbers[0] = 10
print(numbers)
```

```
[10, 2, 3, 4, 5]
```

List Operations

Python provides several built-in methods for working with lists. Some of the most commonly used methods include:

append() - adds an item to the end of the list

insert() - adds an item at a specific position in the list

remove() - removes an item from the list

pop() - removes and returns the last item in the list

clear() - removes all items from the list

index() - returns the index of the first occurrence of a given item in the list

count() - returns the number of occurrences of a given item in the list

sort() - sorts the items in the list in ascending order

reverse() - reverses the order of the items in the list

Examples:

```
numbers.append(6)
print(numbers)
```

```
[10, 2, 3, 4, 5, 6]
```

```
numbers.insert(1, 20)
print(numbers)
```

```
[10, 20, 2, 3, 4, 5, 6]
```

```
numbers.remove(3)
print(numbers)
```

```
[10, 20, 2, 4, 5, 6]
```

```
In [4]:  numbers.append(6)
         print(numbers)

[10, 2, 3, 4, 5, 6]
```

```
In [5]:  numbers.insert(1, 20)
         print(numbers)

[10, 20, 2, 3, 4, 5, 6]
```

```
In [6]:  numbers.remove(3)
         print(numbers)

[10, 20, 2, 4, 5, 6]
```

It is also possible to use list comprehension, which is a concise way to create and manipulate lists.

```
squares = [x**2 for x in range(1, 6)]
print(squares)
```

```
[1, 4, 9, 16, 25]
```

```
In [7]:  squares = [x**2 for x in range(1, 6)]
         print(squares)

[1, 4, 9, 16, 25]
```

Lists are very powerful and flexible, and they are used extensively in various Python libraries and frameworks. They are also very easy to use, making them a great choice for beginners.

Tuples

A tuple is a built-in data structure in Python that is similar to a list, but it is immutable, meaning that its elements cannot be modified once created. A tuple is created by placing a comma-separated sequence of elements inside parentheses. For example, a tuple containing the elements "apple", "banana", and "cherry" can be created as follows:

```
fruits = ("apple", "banana", "cherry")
```

Tuples have several advantages over lists. For example, because they are immutable, they can be used as keys in dictionaries, which is not possible with lists. In addition, tuples are generally faster and use less memory than lists, which can be beneficial when working with large amounts of data.

One of the main differences between lists and tuples is that while lists are mutable, tuples are immutable. This means that you can change the elements of a list but you can't change the elements of a tuple. For example, you can change an element in a list by using the indexing operator, but you can't do that with a tuple.

```
#example of list
fruits = ["apple", "banana", "cherry"]
fruits[1] = "mango"
print(fruits) # ["apple", "mango", "cherry"]
#example of tuple
fruits = ("apple", "banana", "cherry")
fruits[1] = "mango" #TypeError: 'tuple' object does not support item
assignment
```

```
#example of List
fruits = ["apple", "banana", "cherry"]
fruits[1] = "mango"
print(fruits) # ["apple", "mango", "cherry"]
#example of tuple
fruits = ("apple", "banana", "cherry")
fruits[1] = "mango" #TypeError: 'tuple' object does not support item assignment

['apple', 'mango', 'cherry']

---------------------------------------------------------------------------
TypeError                                 Traceback (most recent call last)
Input In [9], in <cell line: 7>()
      5 #example of tuple
      6 fruits = ("apple", "banana", "cherry")
----> 7 fruits[1] = "mango"

TypeError: 'tuple' object does not support item assignment
```

Tuples also have some built-in functions that are not available for lists. For example, the count() function returns the number of occurrences of a given element in a tuple, while the index() function returns the index of the first occurrence of a given element in a tuple.

```
#example
fruits = ("apple", "banana", "cherry", "banana")
print(fruits.count("banana")) # 2
print(fruits.index("cherry")) # 2
```

In general, tuples are best used when you have a fixed set of elements that you don't need to change, while lists are more appropriate when you need to add or remove elements frequently.

Sets

Sets in Python are a built-in data structure that allows you to store unique items in an unordered collection. Sets are often used to check for membership, remove duplicates from a list, or perform mathematical set operations such as union, intersection, and difference.

Creating a set in Python is simple, you can use the set() constructor or use curly braces {} to create an empty set. To add items to a set, you can use the add() method or use the union operator |. Here is an example:

```
# Creating an empty set
my_set = set()
# Adding items to a set
my_set.add(1)
my_set.add(2)
my_set.add(3)
# Creating a set with items
my_set = {1, 2, 3}
# Union operation
set1 = {1, 2, 3}
set2 = {3, 4, 5}
set3 = set1 | set2
print(set3) # Output: {1, 2, 3, 4, 5}
```

You can also use the built-in functions such as len() to check the number of items in a set, in to check for membership, and clear() to remove all items from a set.

Sets in python are unordered collections of unique items, this makes it very useful when you want to remove duplicates from a list or check for membership. It is also efficient in terms of performance as the checking for membership has an average time complexity of O(1).

Dictionaries

Dictionaries in Python are an unordered collection of key-value pairs. They are also known as **associative arrays** or **hash maps**. Dictionaries are defined using curly braces {} and keys and values are separated by a colon (:).

Dictionaries are extremely useful for data manipulation and organization. They allow for fast and efficient access to data using keys instead of indexing. For example, you can use a dictionary to store a list of employee names as keys and their corresponding employee ID as values. This allows you to quickly look up an employee's ID by their name.

Creating a dictionary in python is as simple as:

```
employee_dict = {'John': 1, 'Mary': 2, 'Bob': 3}
```

You can also create an empty dictionary using the dict() constructor:

```
employee_dict = dict()
```

You can access the elements of a dictionary by using the keys. For example, to access the employee ID of John, you can use:

```
employee_dict['John']
```

You can add, update, and delete elements in a dictionary by using the keys. For example, to add a new employee to the dictionary:

```
employee_dict['Jane'] = 4
```

To update an employee's ID:

```
employee_dict['John'] = 5
```

And to delete an employee from the dictionary:

```
del employee_dict['Bob']
```

Dictionaries also have several built-in methods for manipulating and working with the data. Some commonly used methods include:

.keys() : returns a list of all the keys in the dictionary

.values() : returns a list of all the values in the dictionary

.items() : returns a list of all the key-value pairs in the dictionary

.get(key) : returns the value of the specified key, if the key is not found it returns None

.pop(key) : removes the specified key-value pair from the dictionary and returns the value

.clear() : removes all key-value pairs from the dictionary

In summary, dictionaries in python are a powerful and versatile data structure. They allow for efficient data manipulation and organization and have a wide range of built-in methods for working with the data.

Python also has a built-in data structure called array, which is similar to a list but is more efficient for numerical operations.

Choose the right data structure for the task at hand to ensure efficient and effective data manipulation. Understanding the characteristics and capabilities of these built-in data structures will help you make the best use of them in your data analysis projects.

3.2 BUILT-IN FUNCTIONS

Built-in functions are pre-defined functions in Python that are always available and can be used without importing any module. These functions are built into the Python language and can be used to perform a wide range of tasks, from simple mathematical calculations to more complex string and data manipulation. Some examples of built-in functions in Python include:

Mathematical functions

Python, being a high-level programming language, provides a wide range of built-in mathematical functions that can be used for various mathematical operations. Some of the commonly used mathematical functions in Python are:

abs() - returns the absolute value of a number

max() - returns the largest element in an iterable

min() - returns the smallest element in an iterable

round() - rounds a number to the nearest integer or to a specified number of decimal places

pow() - raises a number to a specified power

sqrt() - returns the square root of a number

sum() - returns the sum of all elements in an iterable

math.floor() - rounds a number down to the nearest integer

math.ceil() - rounds a number up to the nearest integer

Here's an example covering some of the math functions in Python:

```
import math

# Absolute value
print(abs(-5)) # Output: 5

# Maximum and minimum values
print(max(2, 5, 1)) # Output: 5
print(min(2, 5, 1)) # Output: 1

# Exponential and logarithmic functions
print(math.exp(2)) # Output: 7.38905609893065
print(math.log(2)) # Output: 0.6931471805599453
print(math.log10(2)) # Output: 0.3010299956639812

# Trigonometric functions
print(math.sin(0)) # Output: 0.0
print(math.cos(math.pi)) # Output: -1.0
print(math.tan(math.pi/4)) # Output: 0.9999999999999999

# Rounding and floor/ceiling functions
print(round(2.345, 2)) # Output: 2.35
print(math.floor(2.345)) # Output: 2
print(math.ceil(2.345)) # Output: 3

# Power and square root functions
print(pow(2, 3)) # Output: 8
print(math.sqrt(16)) # Output: 4.0
```

This program demonstrates the use of various math functions in Python, including absolute value, maximum and minimum values, exponential and logarithmic functions, trigonometric functions, rounding and floor/ceiling functions, and power and square root functions.

It's also possible to use mathematical operators like +, -, *, /, %, ** to perform mathematical operations.

In addition to these built-in functions, Python also provides a module called math which contains additional mathematical functions like trigonometric functions, logarithmic functions, etc.

For example, math.sin() returns the sine of an angle in radians, math.log() returns the logarithm of a number to the base 10 or natural logarithm and so on.

It's important to note that while using mathematical functions, we must make sure that the input is in the correct format and within the valid range of the function.

It's also important to import the math module to use these additional mathematical functions in python.

```
import math
math.sqrt(16) #4.0
math.sin(math.pi/2) #1.0
```

In addition to these, python also provides built-in functions for type casting such as int(), float(), str() etc.

```
int('123') # 123
float(123) # 123.0
```

Python also provides built-in functions for type conversion between different data types such as list(), dict(), set() etc.

```
list((1,2,3)) # [1,2,3]
```

It's also possible to use built-in functions for type checking such as isinstance()

```
isinstance(123, int) # True
```

Overall, python's built-in functions and modules provides a wide range of functionality for mathematical operations and data type conversions, making it easy for developers to perform these operations without having to write additional code.

Type Conversion Functions

In Python, type-conversion functions are used to convert one data type to another. These functions are built-in and can be used on various data types such as integers, floating-point numbers, strings, lists, etc.

Here are some of the most commonly used type-conversion functions in Python:

int() - This function is used to convert any data type to an integer.

For example, int("10") will return 10,

int(10.5) will return 10, and

int([1, 2, 3]) will raise a TypeError.

float() - This function is used to convert any data type to a floating-point number.

For example, float("10.5") will return 10.5,

float(10) will return 10.0, and

float([1, 2, 3]) will raise a TypeError.

str() - This function is used to convert any data type to a string.

For example, str(10) will return "10",

str(10.5) will return "10.5", and

str([1, 2, 3]) will return "[1, 2, 3]".

list() - This function is used to convert any data type to a list.

For example, list("hello") will return ['h', 'e', 'l', 'l', 'o'], list(10) will raise a TypeError, and

list({1: 'a', 2: 'b'}) will return [1, 2].

tuple() - This function is used to convert any data type to a tuple.

For example, tuple("hello") will return ('h', 'e', 'l', 'l', 'o'),

tuple(10) will return (10,), and

tuple([1, 2, 3]) will return (1, 2, 3).

set() - This function is used to convert any data type to a set.

For example, set("hello") will return {'h', 'e', 'l', 'o'},

set(10) will raise a TypeError, and

set([1, 2, 3]) will return {1, 2, 3}.

dict() - This function is used to convert any data type to a dictionary.

For example, dict("hello") will raise a TypeError,

dict(10) will raise a TypeError, and

dict([(1, 'a'), (2, 'b')]) will return {1: 'a', 2: 'b'}.

When using type-conversion functions, it's important to keep in mind that the input data type must be compatible with the desired output data type. Incompatible data types will raise a TypeError. It's also important to consider the context in which the data will be used, as certain data types may be more appropriate for specific tasks.

String functions

In Python, there are a number of built-in functions that can be used to manipulate strings. Some of the most commonly used string manipulation functions include:

len(string): Returns the length of the input string.

str.lower(): Returns a copy of the input string in lowercase.

str.upper(): Returns a copy of the input string in uppercase.

str.strip(): Returns a copy of the input string with leading and trailing white spaces removed.

str.replace(old, new): Returns a copy of the input string with all occurrences of the old string replaced with the new string.

str.split(separator): Returns a list of substrings obtained by splitting the input string at each occurrence of the separator.

str.join(iterable): Returns a string obtained by concatenating the elements of the iterable using the input string as the separator.

str.find(sub): Returns the index of the first occurrence of the substring in the input string, or -1 if the substring is not found.

str.index(sub): Same as str.find(), but raises an error if the substring is not found.

str.count(sub): Returns the number of occurrences of the substring in the input string.

Here is an example program that covers several string manipulation functions in Python:

```
# String Manipulation Example

# Define a string
my_string = "Hello World!"

# Print the original string
print("Original String:", my_string)

# Convert string to uppercase
my_string_upper = my_string.upper()
print("Uppercase String:", my_string_upper)

# Convert string to lowercase
my_string_lower = my_string.lower()
print("Lowercase String:", my_string_lower)

# Capitalize the first letter of the string
my_string_cap = my_string.capitalize()
print("Capitalized String:", my_string_cap)
```

```
# Replace a substring in the string
my_string_replace = my_string.replace("World", "Python")
print("Replaced String:", my_string_replace)

# Split the string into a list of substrings
my_string_split = my_string.split(" ")
print("Split String:", my_string_split)

# Join a list of substrings into a string
my_string_join = "-".join(my_string_split)
print("Joined String:", my_string_join)

# Check if a substring is in the string
if "World" in my_string:
    print("Substring Found!")
else:
    print("Substring Not Found.")
```

The output will look like this:

```
Original String: Hello World!
Uppercase String: HELLO WORLD!
Lowercase String: hello world!
Capitalized String: Hello world!
Replaced String: Hello Python!
Split String: ['Hello', 'World!']
Joined String: Hello-World!
Substring Found!
```

This program defines a string "Hello World!" and performs the following manipulations:

1. Convert the string to uppercase using **upper()**.

2. Convert the string to lowercase using **lower()**.

3. Capitalize the first letter of the string using **capitalize()**.

4. Replace the substring "World" with "Python" using **replace()**.

5. Split the string into a list of substrings using **split()**.

6. Join the substrings in the list using **-** as the separator using **join()**.

7. Check if the substring "World" is in the string using the **in** keyword.

These are just a few examples of the built-in string manipulation functions in Python. There are many other functions that can be used for more advanced manipulation of strings, such as regular expressions and Unicode handling.

List functions

List functions in python are a set of built-in functions that are used to perform various operations on lists. Some of the most commonly used list functions in python are:

len(list): This function returns the number of elements in the given list.

max(list): This function returns the maximum element in the given list.

min(list): This function returns the minimum element in the given list.

list.append(element): This function adds the given element to the end of the list.

list.extend(iterable): This function adds all elements of the given iterable to the end of the list.

list.insert(index, element): This function inserts the given element at the specified index in the list.

list.remove(element): This function removes the first occurrence of the given element in the list.

list.pop(index): This function removes and returns the element at the specified index in the list.

list.index(element): This function returns the index of the first occurrence of the given element in the list.

list.count(element): This function returns the number of occurrences of the given element in the list.

list.sort(): This function sorts the elements of the list in ascending order.

list.reverse(): This function reverses the order of the elements in the list.

list.clear(): This function removes all elements from the list.

Some of these functions, like append, extend and remove change the original list, while others like max, min and index return a value but don't change the original list.

Some of the functions, like sort and reverse are inplace and change the original list, while others like sorted and reversed return a new list without changing the original one.

Example:

```
numbers = [1, 2, 3, 4, 5]

# Using len function
print(len(numbers))
# Output: 5

# Using max function
print(max(numbers))
# Output: 5

# Using min function
print(min(numbers))
# Output: 1

# Using append function
numbers.append(6)
print(numbers)
# Output: [1, 2, 3, 4, 5, 6]

# Using insert function
numbers.insert(0, 0)
print(numbers)
# Output: [0, 1, 2, 3, 4, 5, 6]

# Using remove function
numbers.remove(2)
print(numbers)
# Output: [0, 1, 3, 4, 5, 6]
```

```
# Using pop function
numbers.pop(0)
print(numbers)
# Output: [1, 3, 4, 5, 6]

# Using index function
print(numbers.index(3))
# Output: 1

# Using count function
print(numbers.count(3))
# Output: 1

# Using sort function
numbers.sort()
print(numbers)
# Output: [1, 3, 4, 5, 6]

# Using reverse function
numbers.reverse()
```

Miscellaneous functions

Miscellaneous functions in python are a collection of various functions that do not fit into the category of mathematical functions, type-conversion functions, string manipulation functions, or list functions. Some examples of miscellaneous functions in python include:

zip(): This function is used to combine multiple iterables into one. It takes in any number of iterables as arguments and returns an iterator that produces tuples containing elements from each of the input iterables.

enumerate(): This function is used to iterate over a sequence and return a tuple containing the index and the element at that index. It takes in an iterable as an argument and returns an iterator that produces tuples in the format (index, element).

filter(): This function is used to filter elements from a sequence based on a certain condition. It takes in a function and an iterable as arguments and returns an iterator that produces elements from the input iterable that satisfy the condition specified in the function.

map(): This function is used to apply a certain operation to all elements of a sequence. It takes in a function and an iterable as arguments and returns an iterator that produces the result of applying the function to each element of the input iterable.

reduce(): This function is used to apply a certain operation to all elements of a sequence and reduce it to a single value. It takes in a function and an iterable as arguments and returns the result of applying the function cumulatively to the elements of the input iterable.

These are some examples of miscellaneous functions in python that can be useful in various data manipulation and processing tasks. Python also has built-in modules like itertools and functools which contain more advanced and specialized miscellaneous functions.

In order to use these functions, you simply need to call the function by name and pass in the appropriate arguments. For example, to use the len() function to find the length of a string, you would write len("hello world").

python has a large number of built-in functions, this section covers a few examples, for more information on built-in functions in python you can refer to python documentation.

3.3 ANONYMOUS FUNCTIONS

An anonymous function, also known as a lambda function, is a function that is defined without a name. In Python, anonymous functions are created using the **lambda keyword**. These functions are useful when you need to define a function for a short period of time, or when the function is only used once in the code.

Syntax for creating a lambda function:

lambda arguments: expression

For example, the following lambda function takes two arguments and returns their sum:

```
add = lambda x, y: x + y
print(add(3,4)) # Output: 7
```

Lambda functions can be used wherever a function is expected, such as in the map() and filter() functions.

```
numbers = [1, 2, 3, 4, 5]
squared_numbers = list(map(lambda x: x**2, numbers))
print(squared_numbers) # Output: [1, 4, 9, 16, 25]
```

Lambda functions can only contain one expression and cannot contain statements or annotations. Also, they are limited in terms of functionality compared to regular functions, but they can be useful when you need a small, simple function.

3.4 DEFINING A FUNCTION

In Python, a function is a block of code that performs a specific task. Functions are defined using the "def" keyword, followed by the function name and a pair of parentheses. Inside the parentheses, we can specify the parameters that the function takes. These are also known as function arguments.

When we call a function, we pass in values for these parameters, which are then used inside the function to perform its task. The function can then use these values to perform some operation, and can also return a value to the caller. The value that a function returns is known as the return value.

For example, let's consider a simple function that takes two numbers as arguments and returns their sum:

```
def add_numbers(x, y):
    return x + y
result = add_numbers(3, 4)
print(result)  # Output: 7
```

In this example, the function "add_numbers" takes two arguments "x" and "y" and returns their sum. When we call the function with the values 3 and 4, the function returns 7 which is stored in the variable "result" and printed out.

Values passed to a function as arguments are stored in a different memory location from the variables that are used to call the function. This means that any changes made to the arguments inside the function do not affect the variables used to call the function. This concept is known as "pass by value" in Python.

In Python, we can also define default values for function arguments, so that if the caller does not provide a value for that argument, the default value will be used. This can make it easier to write flexible and reusable functions.

Here's an example program that defines a function with different scenarios:

```
# Define a function that takes two arguments and returns their product
def multiply(a, b):
    return a * b

# Define a function with a default argument
def greet(name="World"):
    print("Hello,", name)

# Define a function with variable number of arguments
def add(*args):
    sum = 0
    for arg in args:
        sum += arg
    return sum

# Define a function with keyword arguments
def details(name, age, city):
    print("Name:", name)
    print("Age:", age)
    print("City:", city)

# Call the functions with different arguments
print(multiply(2, 3))       # Output: 6

greet()                     # Output: Hello, World
greet("John")               # Output: Hello, John

print(add(1, 2, 3))         # Output: 6
print(add(4, 5, 6, 7))      # Output: 22

details(name="John", age=30, city="New York")
# Output:
# Name: John
# Age: 30
```

```
# City: New York
```

In this example, we defined a function **multiply()** that takes two arguments and returns their product. We also defined a function **greet()** with a default argument that prints a greeting message, and a function **add()** with variable number of arguments that calculates their sum. Finally, we defined a function **details()** with keyword arguments that prints some details about a person.

We then called these functions with different arguments to demonstrate their behavior. This example covers all the possible scenarios for defining a function in Python.

3.5 NAMESPACE AND SCOPE OF A FUNCTION

When writing code in any programming language, it's important to understand the concept of namespaces and scope. These terms are closely related and are used to determine the accessibility or visibility of variables, functions, and other resources in a program.

A namespace is a container that holds a set of identifiers. Each identifier is associated with an object, such as a variable or a function. A namespace can be thought of as a dictionary, where the keys are the identifiers and the values are the associated objects. For example, in Python, the built-in functions such as print() and len() are stored in a built-in namespace. Similarly, the variables and functions that you create in your code are stored in a local namespace.

Scope refers to the portion of the code in which an identifier can be accessed. In other words, it determines the visibility of an identifier in a program. There are two types of scope in Python: global and local.

A global scope is the portion of the code where a variable or function can be accessed from anywhere. These variables and functions are defined in the global namespace and are available throughout the entire program.

A local scope, on the other hand, is the portion of the code where a variable or function can only be accessed within a specific block or function. These variables and functions are defined in the local namespace and are only available within that specific block or function.

To understand the concept of namespace and scope of a function in Python, let's take an example:

```
x = 10  # global variable

def my_function(y):
    z = x + y  # x is a global variable, y is a local variable
    return z

result = my_function(5)
print(result)  # Output: 15
```

In this example, we have defined a global variable **x** outside the function **my_function**. Inside the function, we have defined a local variable **y**, which is passed as an argument to the function. We have also defined a local variable **z**, which is the sum of **x** and **y**.

When the function is called with the argument **5**, the value of **z** is computed and returned, which is equal to **x + y**, i.e., **10 + 5 = 15**. The variable **result** is assigned the returned value of the function, which is then printed to the console.

In this example, we can see that the global variable **x** is accessible inside the function, and we can use it in our computation. However, the variables **y** and **z** are local variables, and they are only accessible within the function.

When a variable or function is defined in a local scope, it is said to be shadowed if a variable or function with the same name is defined in the global scope. This means that the global variable or function is no longer accessible within that local scope.

In Python, the **LEGB rule** is used to determine the scope of an identifier. The rule stands for **Local, Enclosing, Global, and Built-in**. The interpreter checks these scopes in the following order: Local, Enclosing, Global, and Built-in. If the interpreter finds a match in the Local scope, it stops searching and uses that match. If it doesn't find a match in the Local scope, it continues to the Enclosing scope, and so on.

Here's an example of the LEGB rule in action:

```
x = 'global'

def outer():
```

```
    x = 'enclosing'

    def inner():
        x = 'local'
        print(x)

    inner()

outer()
print(x)
```

In this example, the variable **x** is defined in three different scopes: global, enclosing, and local. The **outer()** function has an enclosing scope because it encloses the **inner()** function. When the **inner()** function is called, it first looks for the variable **x** in its local scope and finds it, so it prints out the value 'local'. If **x** was not defined in its local scope, it would then look in the enclosing scope of **outer()**, and if it was not defined there, it would look in the global and built-in namespaces.

When the **outer()** function is called, it does not print anything because it does not have a **print** statement. However, it defines the variable **x** in its enclosing scope. Finally, the last **print** statement outside of any function prints out the value of **x** in the global namespace, which is 'global'.

Understanding the LEGB rule is important for avoiding variable name collisions and for understanding how Python looks for and resolves variable names.

3.6 HANDLING FILES IN PYTHON

File handling in Python is a way to work with files and perform various operations such as reading, writing, and manipulating them. Python provides several built-in functions and methods for handling files, including:

open(): This function is used to open a file and return a file object, which can be used to perform various file operations. The basic syntax for opening a file is:

"file_object = open(file_name, mode)"

The file_name parameter is the name of the file to be opened, and the mode parameter is used to specify the mode in which the file should be opened (e.g. 'r' for reading, 'w' for writing, 'a' for appending, etc.).

read(): This method is used to read the contents of a file. The syntax for reading a file is:

"file_object.read(size)"

The size parameter is optional and specifies the number of characters to be read from the file. If size is not specified, the entire contents of the file are read.

write(): This method is used to write content to a file. The syntax for writing to a file is:

"file_object.write(string)"

The string parameter is the content that is written to the file.

close(): This method is used to close a file. The syntax for closing a file is:

"file_object.close()"

tell(): This method returns an integer that represents the current position of the file pointer.

seek(): This method is used to change the current position of the file pointer. The syntax for using the seek method is:

"file_object.seek(offset, from_what)"

The offset parameter is the number of bytes to be moved, and the from_what parameter specifies the reference point for the offset (e.g. 0 for the beginning of the file, 1 for the current position, or 2 for the end of the file).

Here's an example program for handling files in Python:

```
# Open a file for reading
with open('example.txt', 'r') as file:
    # Read the contents of the file
    contents = file.read()

# Open a file for writing
with open('output.txt', 'w') as file:
```

```
    # Write some text to the file
    file.write('This is some output\n')

# Open a file for appending
with open('output.txt', 'a') as file:
    # Append some more text to the file
    file.write('This is some additional output\n')

# Open a file in binary mode
with open('binary.dat', 'wb') as file:
    # Write some binary data to the file
    file.write(b'\x01\x02\x03\x04')

# Open a file in text mode with encoding
with open('unicode.txt', 'w', encoding='utf-8') as file:
    # Write some text with unicode characters to the file
    file.write('This is some Unicode text: \u2603\n')
```

In this program, we use the **open()** function to open files in different modes:

- **r** mode is used for reading a file
- **w** mode is used for writing a new file, overwriting any existing file with the same name
- **a** mode is used for appending to an existing file
- **wb** mode is used for writing binary data to a file
- **w** mode with encoding is used for writing text data with a specific encoding (in this case, UTF-8)

We use the **with** statement to automatically close the file after we're done working with it. This is important to ensure that resources are properly released and that any changes we made to the file are saved.

Inside each **with** block, we use the **file** object to perform various file operations. In the first block, we read the contents of the file using the **read()** method. In the next two blocks, we write text to the file using the **write()** method. In the fourth block, we write binary data to the file. Finally, in the last block, we write text with Unicode characters to the file using the specified encoding.

By using these various modes and methods, we can perform a wide range of file operations in Python. Python also provides several other file-related functions and methods that can be used to perform various file operations.

3.7 EXCEPTION HANDLING

Exception handling in Python is a process of handling errors that may occur during the execution of a program. These errors are known as exceptions. The built-in module, "exceptions" contains several classes that can be used to handle exceptions in Python.

There are two types of exceptions in Python: built-in exceptions and user-defined exceptions.

Built-in exceptions are those that are predefined in Python and are raised when a specific condition occurs. Examples of built-in exceptions include ValueError, TypeError, and KeyError.

User-defined exceptions are those that are created by the user and raised when a specific condition occurs.

The basic structure of exception handling in Python is as follows:

```
try:
# code that may raise an exception
except ExceptionType:
# code to handle the exception
```

In the above code snippet, the code inside the try block is executed. If an exception occurs, the code inside the except block is executed. The ExceptionType specifies the type of exception that the except block is handling.

It's also possible to use the finally block, which will be executed regardless of whether an exception occurred or not.

```
try:
# code that may raise an exception
except ExceptionType:
# code to handle the exception
```

```
finally:
# code that will always be executed
```

Additionally, multiple exception types can be handled by using multiple except blocks.

```
try:
# code that may raise an exception
except ExceptionType1:
# code to handle the exception
except ExceptionType2:
# code to handle the exception
In order to raise an exception, the raise keyword is used.
raise ExceptionType("Error message")
```

It's also possible to raise an exception with an existing exception as the cause, by passing it as the argument of the raise statement.

```
raise ExceptionType("Error message") from original_exception
```

Here's a programming example for exception handling in Python:

```
try:
    # Attempt to open file for reading
    file = open("example.txt", "r")
    # Attempt to read contents of file
    contents = file.read()
    # Close the file
    file.close()
    # Attempt to convert contents to integer
    num = int(contents)
    # Print the result
    print("The number is:", num)
except FileNotFoundError:
    print("File not found.")
except ValueError:
    print("Contents of file are not a valid integer.")
```

```
except Exception as e:
    print("An unexpected error occurred:", e)
finally:
    print("Program execution completed.")
```

In this example, we're trying to open a file called "example.txt", read its contents, convert those contents to an integer, and print the result. However, there are a number of things that could go wrong along the way, such as the file not existing, the contents of the file not being a valid integer, or an unexpected error occurring.

To handle these possibilities, we use a **try** block. Within the **try** block, we attempt to perform the desired operations. If an error occurs, we catch the exception using an **except** block. In this example, we have several different **except** blocks to handle different types of exceptions, such as **FileNotFoundError** and **ValueError**.

We also have a catch-all **except** block at the end, which will catch any exceptions that aren't handled by the previous blocks. This can be useful for debugging, as it allows us to see what went wrong in unexpected situations.

Finally, we have a **finally** block, which is executed regardless of whether an exception occurred. This is a good place to put any code that should be run regardless of the outcome of the **try** block, such as closing files or cleaning up resources.

Using **try/except** blocks is a powerful way to handle errors in your Python code and ensure that your program continues to execute even in the face of unexpected problems.

3.8 DEBUGGING TECHNIQUES

Debugging techniques are methods used to identify and fix errors or bugs in a computer program. These techniques are essential for ensuring that code runs correctly and efficiently.

One common debugging technique is using **print statements**, which allows you to see the values of variables at different points in the code. This can help you identify where an error is occurring and what is causing it.

Another technique is **using a debugger**, which is a tool that allows you to step through the code line by line and see the values of variables at each step. This can help you understand how the code is executing and identify the source of an error.

Another technique is **using assert statements**, which allow you to check that certain conditions are true at specific points in the code. If an assert statement fails, the program will stop executing, allowing you to identify the problem.

Another technique is **using a unit test**, which is a test that checks a small piece of code to ensure that it is functioning correctly. This can help you identify errors early on and ensure that changes to the code do not cause unintended consequences.

Another technique is **using a version control system**, which allows you to track changes to your code over time. This can be useful for identifying when an error was introduced and who was responsible for it.

Finally, it's also important to **take a break**, clear your mind and come back later to the problem. Sometimes, a fresh perspective and a clear mind can help to identify the problem more easily.

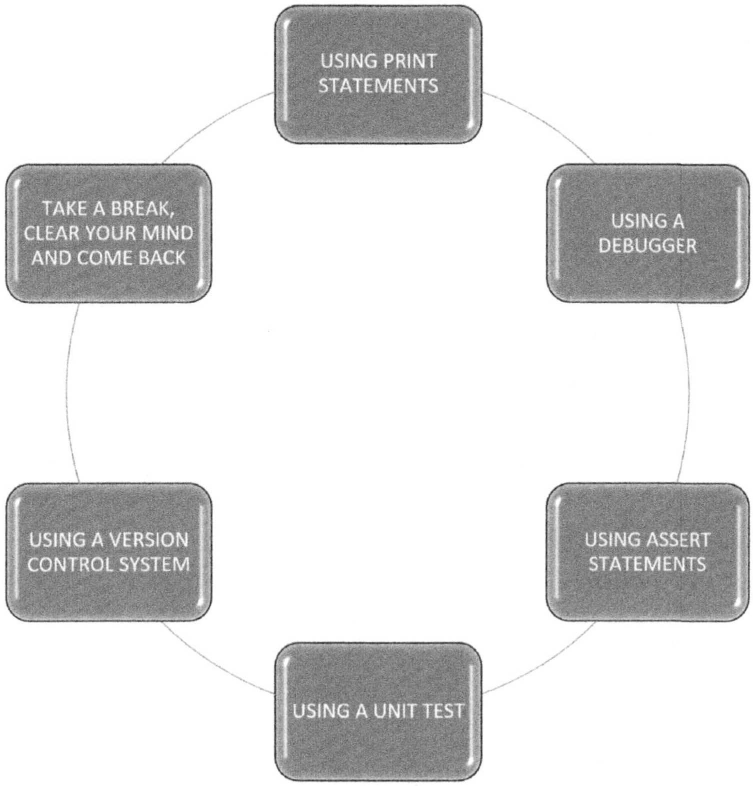

Debugging techniques are an essential part of the programming process, and a good understanding of different techniques can help you to find and fix errors more quickly and efficiently.

3.9 BEST PRACTICES FOR WRITING PYTHON CODE

Python is a versatile and powerful programming language that is widely used in data science, web development, and other fields. However, like any other programming language, it can be easy to make mistakes or write code that is difficult to understand or maintain. In order to write efficient and maintainable code, it is important to follow some best practices. In this section, we will discuss some of the best practices for writing Python code.

I. **Use meaningful variable and function names**: Choosing clear, meaningful variable and function names can make your code much easier to understand and maintain. Avoid using single letter variable names or using abbreviations that are not commonly known.

II. **Use appropriate data types**: Python has a variety of built-in data types to choose from, including integers, floats, strings, and lists. Choosing the appropriate data type for your task will make your code more efficient and easier to understand.

III. **Write modular code**: Break your code into smaller, reusable modules or functions. This will make it easier to understand and test your code, and it will also make it easier to reuse your code in other projects.

IV. **Document your code**: Add comments to your code to explain what it does and how it works. This will make it easier for others to understand and work with your code, and it will also help you to understand your own code in the future.

V. **Avoid hardcoding values**: Instead of hardcoding values, use variables or constants to store values that may need to be changed later. This will make it easier to change your code if needed, and it will also make it more readable.

VI. **Use version control**: Use a version control system such as Git to keep track of changes to your code. This will make it easier to collaborate with others, and it will also make it easier to revert to an earlier version of your code if needed.

VII. **Use libraries and frameworks**: Python has a wide variety of libraries and frameworks that can help you to write code more efficiently and effectively. Learn to use these tools, and use them when appropriate.

VIII. **Test your code**: Test your code thoroughly to make sure that it works correctly. Use automated testing tools to help you to test your code efficiently.

IX. **Keep learning**: Python is a constantly evolving language, so it's important to keep learning and staying up to date with the latest best practices and tools.

By following these best practices, you can write high-quality, maintainable code that is easy to understand, test, and modify. This will help you to be more productive and efficient, and it will also make it easier to collaborate with others.

3.10 SUMMARY

- The chapter covers the basics of built-in data structures in Python, including lists, tuples, sets, and dictionaries.

- It also covers built-in functions in Python, including mathematical functions, type-conversion functions, string manipulation functions, list functions, and anonymous functions.
- The chapter also covers exception handling and file handling in Python, and how to handle errors and read/write to files.
- Additionally, the chapter covers important concepts in function writing such as namespace, scope, arguments, parameters, and return values.
- The chapter concludes with coverage of debugging techniques and best practices for writing Python code to improve readability and maintainability.
- e also discussed anonymous functions, also known as lambda functions, and their use cases
- The chapter also covered exception handling in Python and the different types of errors that can occur in a program
- We also discussed file handling in Python, including opening, reading, writing, and closing files

3.11 TEST YOURR KNOWLEDGE

I. What is the difference between a list and a tuple in Python?

a) Lists are mutable, while tuples are immutable
b) Lists can contain any data type, while tuples can only contain numbers
c) Lists use square brackets, while tuples use parentheses
d) Lists are ordered, while tuples are not

II. What is the purpose of the set() function in Python?

a) To create a new set data type
b) To convert a list to a set
c) To find the unique elements in a list or tuple
d) To remove duplicates from a list or tuple

III. What is the purpose of the dict() function in Python?

a) To create a new dictionary data type
b) To convert a list to a dictionary
c) To find the keys and values of a dictionary
d) To remove elements from a dictionary

IV. What is the purpose of the round() function in Python?

a) To round a decimal number to the nearest whole number
b) To round a decimal number to a specified number of decimal places
c) To convert a number to a string
d) To perform addition on a list of numbers

V. **What is the purpose of the len() function in Python?**
a) To find the length of a string
b) To find the length of a list or tuple
c) To find the length of a dictionary
d) All of the above

VI. **What is the purpose of the zip() function in Python?**
a) To combine two or more lists into a single list of tuples
b) To unpack a list of tuples into separate lists
c) To find the common elements between two lists
d) To sort a list of numbers

VII. **What is the purpose of the filter() function in Python?**
a) To filter a list based on a specified condition
b) To remove elements from a list
c) To return the first element of a list
d) To sort a list in ascending or descending order

VIII. **What is the purpose of the map() function in Python?**
a) To apply a specified function to each element of a list or tuple
b) To create a new list based on a specified condition
c) To find the minimum or maximum value in a list
d) To reverse the order of elements in a list

IX. **What is the purpose of the reduce() function in Python?**
a) To apply a specified function cumulatively to the elements of a list or tuple
b) To remove elements from a list or tuple
c) To sort a list or tuple
d) To return the last element of a list or tuple

X. **What is the purpose of the open() function in Python?**
a) To open a file for reading or writing
b) To create a new file
c) To close a file
d) To delete a file

3.12 ANSWERS

I. Answer: a) Lists are mutable, while tuples are immutable
II. Answer: a) To create a new set data type
III. Answer: a) To create a new dictionary data type
IV. Answer: b) To round a decimal number to a specified number of decimal places
V. Answer: d) All of the above
VI. Answer: a) To combine two or more lists into a single list of tuples
VII. Answer: a) To filter a list based on a specified condition
VIII. Answer: a) To apply a specified function to each element of a list or tuple
IX. Answer: a) To apply a specified function cumulatively to the elements of a list or tuple
X. Answer: a) To open a file for reading or writing

04

4 DATA WRANGLING

Data Wrangling is an essential step in the data analysis process. It is the process of transforming and cleaning raw data into a format that can be easily analyzed. It is a crucial step in the data analysis process, as it allows you to clean, transform, and reshape your data so that it can be used for further analysis. In this chapter, we will cover the various techniques and tools available for data wrangling using Python. We will start by acquiring data from various sources, such as CSV files, Excel spreadsheets, and databases. Then, we will learn about the different data cleaning techniques such as handling missing data, removing duplicates, and outlier detection. We will also cover data transformation techniques such as pivoting and reshaping the data. Finally, we will learn about data validation and how to check the data for errors and inconsistencies. By the end of this chapter, you will have a solid understanding of the data wrangling process and the tools available to make it easy to work with data in Python.

4.1 INTRODUCTION TO DATA WRANGLING

Data wrangling, also known as **data munging**, is the process of transforming and mapping data from one format or structure to another in order to make it ready for analysis. It is a critical step in the data analysis process, as it allows you to clean, transform, and reshape your data so that it can be used for further analysis.

The process of data wrangling can be broken down into several steps: data acquisition, data cleaning, data transformation, and data validation. Data acquisition is the process of acquiring data from various sources, such as files, databases, or web scraping. Data cleaning involves removing or correcting errors and inconsistencies in the data, such as missing values, duplicate records, or outliers. Data transformation involves changing the format or structure of the data, such as pivoting or aggregating, in order to make it more suitable for analysis. Data validation is the process of checking the data for errors and inconsistencies, such as duplicate values or missing data.

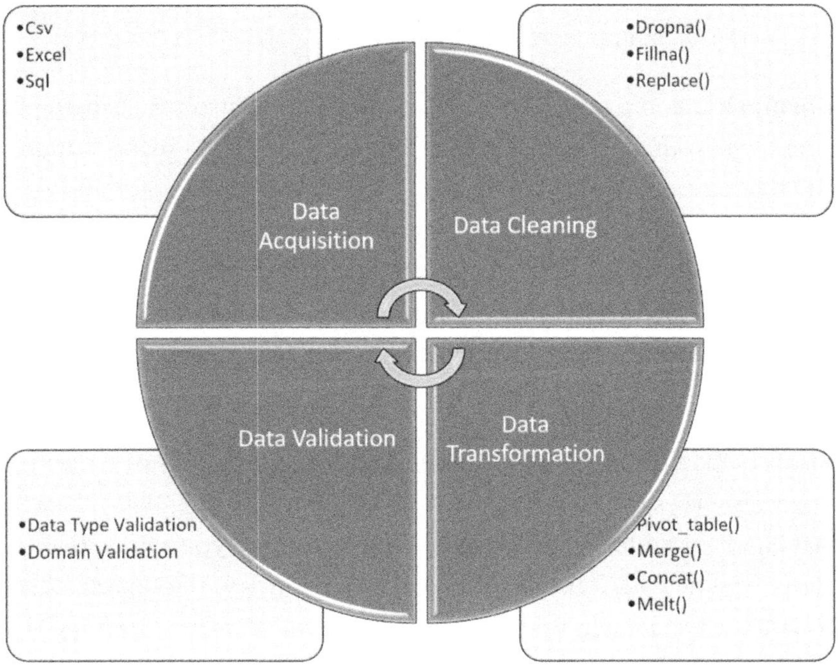

Data acquisition is the first step in data wrangling. It involves acquiring data from various sources, such as CSV files, Excel spreadsheets, and databases. Python provides several libraries that make it easy to work with different data formats and sources, such as Pandas and SQLAlchemy. Pandas is a library that provides fast and flexible data structures for data analysis, such as the DataFrame and Series. It also has powerful data import and export capabilities, making it easy to work with data in a variety of formats, including CSV, Excel, and SQL. SQLAlchemy is a library that provides a consistent interface to various databases, making it easy to work with data stored in databases.

Once the data is acquired, the next step is **data cleaning**. Data cleaning is a crucial step in data wrangling and it involves removing or correcting errors and inconsistencies in the data, such as missing values, duplicate records, or outliers. There are several techniques that can be used to clean data, such as imputation, which is used to fill in missing values; and outlier detection, which is used to identify and remove outliers. Pandas library provides several built-in functions to perform data cleaning tasks such as dropna(), fillna(), and replace() to handle missing values and duplicated() to detect and remove duplicate records.

Data transformation is the next step in data wrangling. It involves changing the format or structure of the data, such as pivoting or aggregating, in order to make it more suitable for analysis. Pivot tables are a powerful tool for data transformation, and they allow you to reshape your data by grouping it by one or more columns and aggregating it by a different column. Pandas provides the pivot_table() function, which can be used to create pivot tables. In addition to pivot tables, data transformation also includes other techniques such as merging, concatenating and reshaping the data using functions like merge(), concat() and melt() respectively.

Data validation is the final step in data wrangling. It is the process of checking the data for errors and inconsistencies, such as duplicate values or missing data. There are several techniques that can be used for data validation, such as **data type validation** and **domain validation**. Data type validation checks that the data conforms to the expected data types, such as integers or dates, while domain validation checks that the data falls within a specified range or set of values. Data validation can be done using various libraries such as Pandas, Numpy and Scipy.

Data wrangling is an iterative process, and it may require several rounds of cleaning and transformation before the data is ready for analysis. Additionally, data wrangling is not a one-time task, and it's important to keep monitoring and maintaining the data throughout the analysis process.

4.2 DATA CLEANING

Data cleaning is a crucial step in data wrangling, and it involves removing or correcting errors and inconsistencies in the data, such as missing values, duplicate records, or outliers. In this section, we will focus on three specific aspects of data cleaning: handling missing data, removing duplicates, and outlier detection.

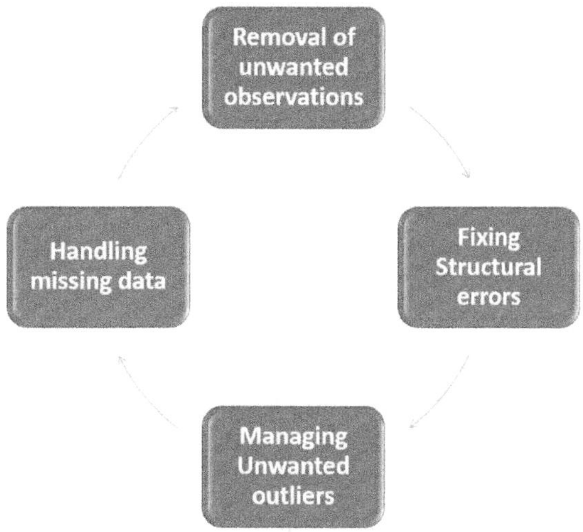

Handling Missing Data

Handling missing data is one of the most common tasks in data cleaning. Missing data can occur for a variety of reasons, such as data entry errors, measurement errors, or non-response. Missing data can have a significant impact on the accuracy and reliability of the analysis, so it's important to handle it correctly. There are several techniques for handling missing data, such as **imputation, deletion,** and **weighting**.

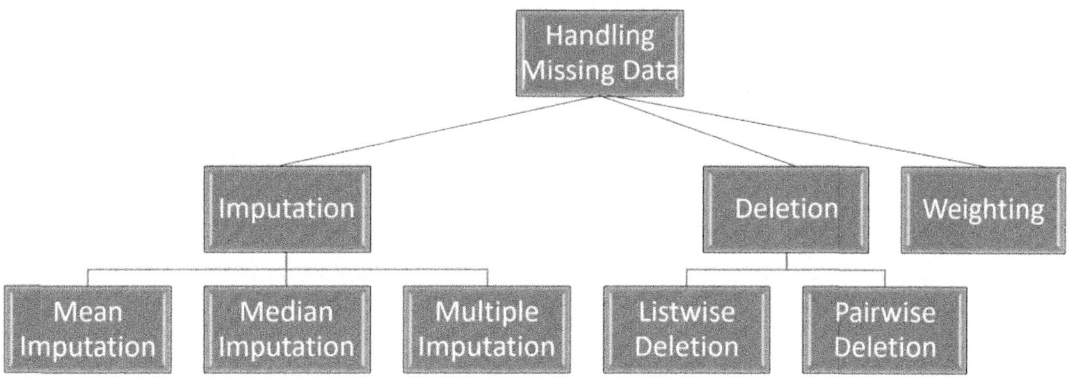

Imputation is the process of replacing missing values with estimated values. There are several imputation methods, such as mean imputation, median imputation, and multiple imputation.

Mean imputation replaces missing values with the mean of the available values, median imputation replaces missing values with the median of the available values, and multiple imputation creates multiple datasets with different imputed values and then combines the results.

Deletion is the process of removing observations or variables with missing values. There are two types of deletion: **listwise deletion** and **pairwise deletion**. Listwise deletion removes an observation if any value is missing, while pairwise deletion only removes the missing values in the analysis.

Weighting is the process of adjusting the analysis to account for missing data. This can be done by weighting the observations or variables with missing values.

In terms of handling missing data, it's important to choose the appropriate imputation method based on the characteristics of the data and the purpose of the analysis. For example, mean imputation is suitable for data with a normal distribution, while multiple imputation is more appropriate for data with a complex distribution. Deletion and weighting are also useful techniques for handling missing data, but they should be used with caution as they may result in a loss of information.

Removing Duplicates

Removing duplicates is another important step in data cleaning. Duplicate records can occur for a variety of reasons, such as data entry errors, data merging errors, or data scraping errors. Duplicate records can have a significant impact on the accuracy and reliability of the analysis, so it's important to remove them. In pandas, the drop_duplicates() function can be used to remove duplicates.

When it comes to removing duplicates, it's important to check the data for duplicates before and after cleaning the data, to ensure that all duplicates have been removed. It's also important to review the duplicates that have been removed to ensure that they are indeed duplicates and not unique records.

Outlier Detection

Outlier detection is the process of identifying and removing outliers in the data. Outliers are data points that are significantly different from the other data points. Outliers can have a significant impact on the accuracy and reliability of the analysis, so it's important to identify and handle them correctly. There are several techniques for outlier detection, such as the Z-score method, the modified Z-score method, and the interquartile range method.

Outlier detection requires a good understanding of the data and the data analysis process. It's important to use appropriate methods for outlier detection based on the characteristics of the data. For example, the Z-score method is suitable for data with a normal distribution, while the interquartile range method is more appropriate for data with a non-normal distribution. It's also important to review the outliers that have been removed to ensure that they are indeed outliers and not valid data points.

In conclusion, data cleaning is a crucial step in data wrangling and it involves removing or correcting errors and inconsistencies in the data. Handling missing data, removing duplicates, and outlier detection are three important aspects of data cleaning. Each of these tasks requires careful attention to detail in order to ensure that the data is cleaned correctly. There are several techniques and tools available in python to handle missing data, remove duplicates and detect outliers.

It's important to have a good understanding of the data and the data analysis process before starting data cleaning, and to keep monitoring and maintaining the data throughout the analysis process.

Keep in mind that data cleaning is not a one-time task, and it's important to keep monitoring and maintaining the data throughout the analysis process. For example, if new data is added to the dataset, it's important to check for missing values, duplicates and outliers again.

Document the data cleaning process, including any decisions made about handling missing data, removing duplicates, and detecting outliers. This documentation will be useful for others who may be working with the data, as well as for future reference.

4.3 DATA TRANSFORMATION AND RESHAPING

Data transformation is the process of changing the format or structure of the data, such as pivoting or aggregating, in order to make it more suitable for analysis. It is an important step in data wrangling, as it allows you to reshape your data into a format that can be easily analyzed. In this section, we will cover the various techniques and tools available for data transformation and reshaping using Python.

If you haven't understood anything, don't worry at this stage, just go through the content, we will explain all these concepts again in later chapter when we discuss Python library.

Reshaping the Data

One of the most common data transformation tasks is reshaping the data. This can be done using various techniques such as pivoting, melting, and stacking.

Pivoting

Pivoting is the process of transforming the data from a long format to a wide format, or vice versa. For example, if you have a dataset with columns for year, month, and revenue, you could pivot the data to have a column for each month and revenue as the values. This can be done using the pivot_table() function in Pandas.

Year	Month	Revenue
2000	January	1000
2001	February	5000
2003	April	9000
2000	January	13000
2001	February	17000
2003	February	21000
2000	April	25000
2001	January	29000
2003	February	33000
2000	April	37000
2001	April	41000
2003	January	45000

Pivotting: Sum of Revenue by month

Year	January	February	April
2000	14000		62000
2001	29000	22000	41000
2003	45000	54000	9000

Melting the Data

Melting is the process of transforming the data from a wide format to a long format. For example, if you have a dataset with columns for each month and revenue as the values, you could melt the data to have a column for year, month, and revenue. This can be done using the melt() function in Pandas.

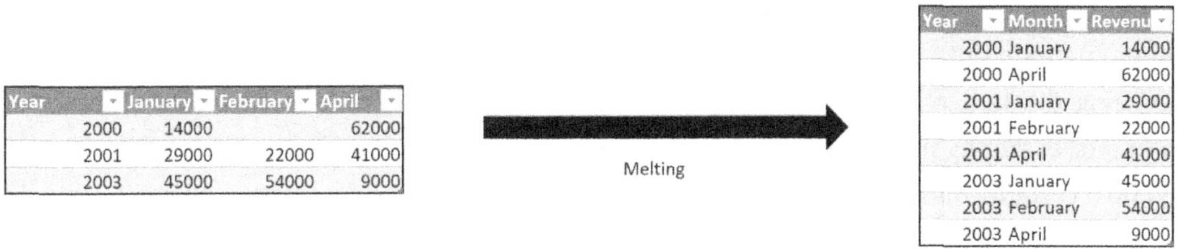

Year	January	February	April
2000	14000		62000
2001	29000	22000	41000
2003	45000	54000	9000

Year	Month	Revenu
2000	January	14000
2000	April	62000
2001	January	29000
2001	February	22000
2001	April	41000
2003	January	45000
2003	February	54000
2003	April	9000

Stacking

Stacking is the process of pivoting the columns to rows, and can be done using the stack() function in pandas. It is useful when you have multi-level columns in your dataframe and you want to reshape it to single level columns.

Aggregating the Data

Another common data transformation task is aggregating the data. This can be done using various techniques such as groupby() and pivot_table(). The groupby() function in Pandas allows you to group the data by one or more columns and apply an aggregate function to another column, such as sum or count. The pivot_table() function allows you to create pivot tables, which are similar to groupby() but allow you to group the data by more than one column and apply an aggregate function to more than one column.

Data Filtering

Data filtering is the process of selecting specific rows or columns from a dataset based on certain criteria. It is an important step in data wrangling, as it allows you to focus on a subset of the data that is relevant to your analysis. In this section, we will cover the various techniques and tools available for data filtering using Python.

One of the most common data filtering tasks is selecting specific rows or columns from a dataset. This can be done using the indexing and slicing techniques in Python. Indexing allows you to select specific rows or columns from a dataset by their position, while slicing allows you to select a range of rows or columns from a dataset. For example, in a pandas DataFrame, you can use the .iloc[] function to access a specific row or column by its position, and the .loc[] function to access a specific row or column by its label.

Another common data filtering task is filtering the data based on certain conditions. This can be done using the logical operators (e.g. >, <, ==, etc.) and the boolean indexing technique in Python. For example, in a pandas DataFrame, you can use the following syntax to filter the data based on a certain condition:

```
df[df['column_name'] > value]
```

It's also possible to filter data based on multiple conditions by using the "&" and "|" operators to combine multiple conditions. For example, in a pandas DataFrame, you can use the following syntax to filter the data based on multiple conditions:

```
df[(df['column_name'] > value) & (df['column_name'] < value)]
```

In addition to the above techniques, one can also filter data based on values of multiple columns, for example

```
df[(df['column_name_1'] > value) & (df['column_name_2'] == value)]
```

Another popular way of filtering data is by using the query function in pandas. The query function allows you to filter data using a string that represents the conditions. This is useful when the filtering conditions are complex and multiple conditions need to be combined. For example,

```
df.query('column_name > value and column_name < value')
```

It's important to note that data filtering is an iterative process, and it may require several rounds of filtering before the data is ready for analysis. Additionally, data filtering is not a one-time task, and it's important to keep monitoring and maintaining the data throughout the analysis process.

Python provides several libraries such as pandas, NumPy, and Matplotlib that make it easy to work with data in Python. The process of data filtering can be broken down into several steps: selecting specific rows or columns, filtering data based on certain conditions, and filtering data based on multiple conditions. Each step is important and requires careful attention to detail in order to ensure that the data is filtered correctly.

Keep in mind that data filtering is not just about removing unwanted data, but also about selecting the relevant data that can be used for further analysis. By filtering the data, you can focus on specific aspects of the data that are relevant to your analysis and make the data more manageable. This makes it easier to work with, and it also helps to reduce the risk of errors in the analysis.

Data Sorting

Data sorting is the process of arranging data in a specific order. It is an important step in data wrangling, as it allows you to organize the data in a way that makes it more meaningful and easier to understand. In this section, we will cover the various techniques and tools available for data sorting using Python.

The most common way to sort data in Python is by using the **sort_values()** function in Pandas. This function allows you to sort a DataFrame or Series by one or more columns. For example, in a pandas DataFrame, you can use the following syntax to sort the data by a specific column:

```
df.sort_values(by='column_name')
```

By default, the data will be sorted in ascending order, but you can also specify to sort in descending order by using the ascending parameter:

```
df.sort_values(by='column_name', ascending=False)
```

Another way to sort data in Python is by using the **sort_index()** function. This function allows you to sort a DataFrame or Series by its index. For example, in a pandas DataFrame, you can use the following syntax to sort the data by its index:

```
df.sort_index()
```

Like sort_values(), sort_index() also accepts an ascending parameter, which can be set to True or False to specify the sort order.

It's also possible to sort data based on multiple columns by passing a list of column names to the sort_values() function. For example, in a pandas DataFrame, you can use the following syntax to sort the data by multiple columns:

```
df.sort_values(by=['column_name_1', 'column_name_2'])
```

In addition to sorting the data based on columns or index, it's also possible to sort the data based on a specific condition, using the sort_values() function in combination with the query() function. For example, in a pandas DataFrame, you can use the following syntax to sort the data based on a specific condition:

```
df.query('column_name > value').sort_values(by='column_name')
```

An important aspect of data sorting is sorting in place. By default, the sorting functions in pandas return a new DataFrame or Series sorted by the specified column(s) or index, but you can also sort the data in place by using the inplace parameter. For example,
df.sort_values(by='column_name', inplace=True)
will sort the DataFrame in place, without creating a new DataFrame.

Here's an example of data sorting using pandas with different scenarios:

```
import pandas as pd

# Creating a sample dataframe
data = {'Name': ['John', 'Sarah', 'Emily', 'Michael', 'Alex'],
        'Age': [25, 30, 18, 42, 28],
        'Salary': [50000, 80000, 35000, 120000, 60000]}
df = pd.DataFrame(data)

# Sorting by a single column
df.sort_values(by='Age', inplace=True)
print("Sorting by Age:\n", df)
```

```
# Sorting by multiple columns
df.sort_values(by=['Age', 'Salary'], inplace=True)
print("\nSorting by Age and Salary:\n", df)

# Sorting in descending order
df.sort_values(by='Salary', ascending=False, inplace=True)
print("\nSorting by Salary in descending order:\n", df)

# Sorting while ignoring the index
df.sort_values(by='Age', ignore_index=True, inplace=True)
print("\nSorting by Age with reset index:\n", df)
```

The output will look like this:

```
Sorting by Age:
      Name  Age  Salary
2    Emily   18   35000
0     John   25   50000
4     Alex   28   60000
1    Sarah   30   80000
3  Michael   42  120000

Sorting by Age and Salary:
      Name  Age  Salary
2    Emily   18   35000
0     John   25   50000
4     Alex   28   60000
1    Sarah   30   80000
3  Michael   42  120000

Sorting by Salary in descending order:
      Name  Age  Salary
3  Michael   42  120000
1    Sarah   30   80000
4     Alex   28   60000
0     John   25   50000
2    Emily   18   35000

Sorting by Age with reset index:
      Name  Age  Salary
0    Emily   18   35000
1     John   25   50000
2     Alex   28   60000
3    Sarah   30   80000
4  Michael   42  120000
```

In the above code, we have created a sample dataframe containing information about employees' names, age, and salary. We then used the **sort_values()** function of pandas to sort the data based on different scenarios. Here's what each scenario represents:

- Sorting by a single column: We sorted the data by the "Age" column by passing "Age" as an argument to the **by** parameter of the **sort_values()** function.
- Sorting by multiple columns: We sorted the data first by "Age" and then by "Salary" by passing both columns as a list to the **by** parameter.
- Sorting in descending order: We sorted the data by "Salary" in descending order by passing **False** to the **ascending** parameter.
- Sorting while ignoring the index: We sorted the data by "Age" and then reset the index by passing **True** to the **ignore_index** parameter.

By using different parameters of the **sort_values()** function, we can easily sort the data based on our requirements in pandas.

Data Grouping

Data grouping is the process of dividing the data into groups based on certain criteria. It is an important step in data wrangling, as it allows you to analyze and understand patterns within the data. In this section, we will cover the various techniques and tools available for data grouping using Python.

One of the most common ways to group data in Python is by using the **groupby()** function in Pandas. This function allows you to group a DataFrame or Series by one or more columns. For example, in a pandas DataFrame, you can use the following syntax to group the data by a specific column:

df.groupby('column_name')

This will return a groupby object, which can be further used to perform various aggregate functions such as sum(), count() and mean().

Another way to group data in Python is by using the **pivot_table()** function. This function allows you to create pivot tables, which are similar to groupby() but allow you to group the data by more than one column and apply an aggregate function to more than one column. For example, in a pandas DataFrame, you can use the following syntax to create a pivot table:

```
df.pivot_table(index='column_name_1', columns='column_name_2',
values='column_name_3', aggfunc='mean')
```

It's also possible to group data based on specific conditions by using the groupby() function in combination with the query() function. For example, in a pandas DataFrame, you can use the following syntax to group the data based on specific conditions:

```
df.query('column_name > value').groupby('column_name_2').mean()
```

Here's an example of data grouping using pandas in Python:

```
import pandas as pd

# create a sample dataframe
data = {'Name': ['Alice', 'Bob', 'Charlie', 'David', 'Eric', 'Frank'],
        'Gender': ['F', 'M', 'M', 'M', 'M', 'M'],
        'Age': [25, 32, 18, 47, 31, 22],
        'Country': ['USA', 'Canada', 'USA', 'USA', 'Canada',
'Canada'],
        'Salary': [50000, 65000, 30000, 80000, 70000, 40000]}

df = pd.DataFrame(data)

print("DataFrame before grouping the data: ")
print(df)

# group data by country and calculate mean salary and age
grouped_data = df.groupby(['Country'])[['Salary', 'Age']].mean()

print("\n\n\n\nDataFrame after grouping the data: ")
print(grouped_data)
```

In this example, we first import the pandas library and create a sample dataframe with information about people, including their name, gender, age, country, and salary. We then use the **groupby** function to group the data by country. The double brackets **[['Salary', 'Age']]** in the **groupby** function are used to specify the columns that we want to group by.

Finally, we use the **mean** function to calculate the average salary and age for each country. The resulting output shows the average salary and age for people in each country.

The output will look like this:

```
DataFrame before grouping the data:
      Name Gender  Age Country  Salary
0    Alice      F   25     USA   50000
1      Bob      M   32  Canada   65000
2  Charlie      M   18     USA   30000
3    David      M   47     USA   80000
4     Eric      M   31  Canada   70000
5    Frank      M   22  Canada   40000
```

```
DataFrame after grouping the data:
               Salary        Age
Country
Canada   58333.333333  28.333333
USA      53333.333333  30.000000
```

In this example, we grouped the data by only one column, but we could also group by multiple columns by passing a list of column names to the **groupby** function. Additionally, we could apply other functions to the grouped data, such as **sum**, **count**, **min**, **max**, and **std**. Data grouping is a powerful technique in data analysis that can help us better understand patterns in our data.

Data transformation also includes other techniques such as merging, concatenating and reshaping the data using functions like merge(), concat() and melt() respectively. Merging is used to combine data from different tables or dataframes, Concatenating is used to combine data from same tables or dataframes and reshaping is used to change the shape of the data.

4.4 DATA VALIDATION

Data validation is the process of checking the data for errors and inconsistencies, such as duplicate values or missing data. It is an important step in data wrangling, as it ensures that the data is accurate, reliable, and fit for purpose. In this section, we will cover the various techniques and tools available for data validation using Python.

One of the most common ways to validate data in Python is by using the **isnull()** or **isna()** function in Pandas. These functions allow you to check for missing data in a DataFrame or Series and return a Boolean value indicating whether the data is missing or not. For example, in a pandas DataFrame, you can use the following syntax to check for missing data in a specific column:

df['column_name'].isnull()

Another way to validate data in Python is by using the **duplicated()** function in Pandas. This function allows you to check for duplicate data in a DataFrame or Series and return a boolean value indicating whether the data is a duplicate or not. For example, in a pandas DataFrame, you can use the following syntax to check for duplicate data in a specific column:

df['column_name'].duplicated()

Here's an example of data validation using the pandas functions **isnull()**, **isna()**, and **duplicated()**. These functions are useful for detecting missing values and duplicates in a pandas DataFrame.

```
import pandas as pd

# Create a sample DataFrame with missing and duplicate values
data = {
    'Name': ['Alice', 'Bob', 'Charlie', 'Dave', 'Eve', 'Alice'],
    'Age': [25, 30, None, 35, 40, 25],
    'Salary': [50000, 60000, 70000, None, 80000, 50000]
}
df = pd.DataFrame(data)

print("DataFrame before validating the data: ")
print(df)

# Check for missing values
print("\n Check for missing values: ")
print(df.isnull())  # Returns a DataFrame with True where there are
missing values

# Check for duplicates
print("\nCheck for duplicates: ")
print(df.duplicated())  # Returns a Series with True where there are
duplicates

# Drop duplicates
df = df.drop_duplicates()

# Fill missing values with mean of column
df = df.fillna(df.mean())

# Check for missing values again
print("\n Check for missing values again after filling missing values:
")
print(df.isnull())
```

The output will look like this:

```
DataFrame before validating the data:
```

	Name	Age	Salary
0	Alice	25.0	50000.0
1	Bob	30.0	60000.0
2	Charlie	NaN	70000.0
3	Dave	35.0	NaN
4	Eve	40.0	80000.0
5	Alice	25.0	50000.0

```
Check for missing values:
```

	Name	Age	Salary
0	False	False	False
1	False	False	False
2	False	True	False
3	False	False	True
4	False	False	False
5	False	False	False

```
Check for duplicates:

0     False
1     False
2     False
3     False
4     False
5      True
dtype: bool
```

```
Check for missing values again after filling missing values:
```

	Name	Age	Salary
0	False	False	False
1	False	False	False
2	False	False	False
3	False	False	False
4	False	False	False

In this example, we create a DataFrame with missing and duplicate values. We then use the **isnull()** function to check for missing values in the DataFrame, which returns a DataFrame with True where there are missing values. We also use the **duplicated()** function to check for duplicates in the DataFrame, which returns a Series with True where there are duplicates.

We then use the **drop_duplicates()** function to drop the duplicate rows in the DataFrame. We also use the **fillna()** function to fill the missing values with the mean of the column.

Finally, we use the **isnull()** function again to check for missing values in the DataFrame after filling the missing values.

Data validation can also be done by using various statistical methods to identify outliers in the data, such as the Z-score method or the Interquartile Range (IQR) method. These methods can be implemented using libraries such as NumPy and Scipy.

Let's discuss two classification of data validation below:

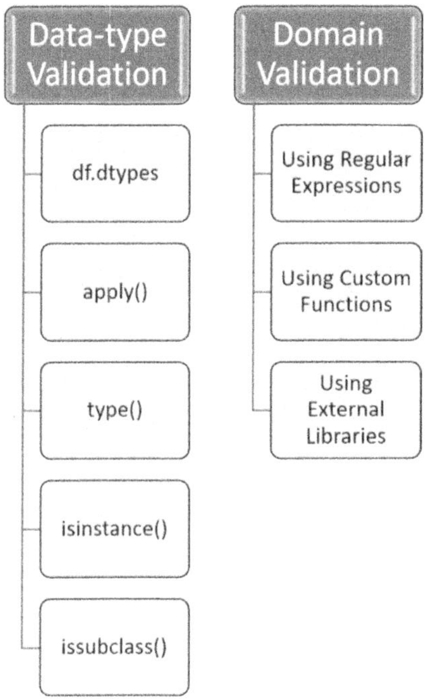

Data-type Validation

Data type validation is the process of checking the data for errors and inconsistencies in terms of data types, such as integer, float, and string. It is an important step in data wrangling, as it ensures that the data is in the correct format for the intended analysis or operation. In this section, we will cover the various techniques and tools available for data type validation using Python.

One of the most common ways to validate data types in Python is by using the **dtypes** property in Pandas. This property allows you to check the data types of all columns in a DataFrame or Series. For example, in a pandas DataFrame, you can use the following syntax to check the data types of all columns:

df.dtypes

Another way to validate data types in Python is by using the **apply()** function in Pandas. This function allows you to apply a specific function to all elements of a DataFrame or Series and check the data types of the elements. For example, in a pandas DataFrame, you can use the following syntax to check the data types of a specific column:

df['column_name'].apply(type)

Data type validation can also be done using built-in Python functions such as **type(), isinstance() and issubclass().** These functions can be used to check the data types of variables and values in a more explicit way.

Here is a coding example for data type validation using pandas:

```
import pandas as pd

# create a sample data frame
df = pd.DataFrame({
    'Name': ['John', 'Jane', 'Adam', 'Ava'],
    'Age': [25, 30, 28, 24],
    'Salary': [50000.00, 60000.00, 55000.00, 45000.00]
})

display(df)

# check data types of columns
print(df.dtypes)

# define a function to check data type of a value
def check_datatype(value):
    if isinstance(value, str):
        return 'string'
    elif isinstance(value, int):
        return 'integer'
    elif isinstance(value, float):
        return 'float'
```

```
    else:
        return 'other'

# apply function to each column of data frame
for col in df.columns:
    print(col, df[col].apply(check_datatype))

# check data type of a single value
value = 'Hello'
print(type(value))
print(isinstance(value, str))
print(issubclass(str, object))
```

The output will look like this:

	Name	Age	Salary
0	John	25	50000.0
1	Jane	30	60000.0
2	Adam	28	55000.0
3	Ava	24	45000.0

```
Name        object
Age          int64
Salary     float64
dtype: object
Name 0    string
1     string
2     string
3     string
Name: Name, dtype: object
Age 0    integer
1     integer
2     integer
3     integer
Name: Age, dtype: object
Salary 0    float
1     float
2     float
3     float
Name: Salary, dtype: object
<class 'str'>
True
True
```

Explanation:

1. Firstly, we import the pandas module and create a sample data frame containing three columns: 'Name', 'Age', and 'Salary'.
2. Then, we use the **dtypes** attribute of the data frame to check the data types of each column. This returns a Series object containing the data types of each column.
3. Next, we define a function **check_datatype** that takes a value as input and checks its data type using the **isinstance** function. If the value is a string, integer, or float, the

function returns the corresponding string 'string', 'integer', or 'float'. Otherwise, it returns 'other'.

4. We then apply the **check_datatype** function to each column of the data frame using the **apply** function. This returns a new data frame where each value has been replaced by its data type.

5. Finally, we check the data type of a single value 'Hello' using the **type**, **isinstance**, and **issubclass** functions. The **type** function returns the actual data type of the value, while the **isinstance** function checks if the value is an instance of a certain class (in this case, **str**). The **issubclass** function checks if **str** is a subclass of **object**.

The process of data type validation can be broken down into several steps: checking the data types of all columns, applying a specific function to check the data types of elements, and using built-in Python functions to check the data types of variables and values.

Domain Validation

Domain validation is the process of checking the data for errors and inconsistencies in terms of domain-specific knowledge or business rules. It is an important step in data wrangling, as it ensures that the data is in accordance with the specific domain or business requirements. In this section, we will cover the various techniques and tools available for domain validation using Python.

Using Regular Expressions

One of the most common ways to validate data based on domain knowledge is by **using regular expressions**. Regular expressions are a powerful tool that allows you to search for patterns within text, and can be used to validate data such as phone numbers, email addresses, or zip codes. In Python, regular expressions can be implemented using the **re** module.

Here's an example of domain validation for phone numbers, email addresses, zip codes, and website using regular expressions in Python:

```
import re

# Phone number validation
def validate_phone_number(number):
    pattern = re.compile(r'^\d{3}-\d{3}-\d{4}$')
```

```
    return bool(pattern.match(number))

# Email address validation
def validate_email(email):
    pattern = re.compile(r'^[\w-]+@[a-zA-Z_]+?\.[a-zA-Z]{2,3}$')
    return bool(pattern.match(email))

# Zip code validation
def validate_zip_code(zip_code):
    pattern = re.compile(r'^\d{5}(?:[-\s]\d{4})?$')
    return bool(pattern.match(zip_code))

# Website validation
def validate_website(url):
    pattern = re.compile(r'^https?://(?:[-\w]+\.)?([-\w]+)\.[a-
z]{2,6}(?:/[-\w\s./?%&=]*)?$')
    return bool(pattern.match(url))

# Testing the validation functions
phone_number = '123-456-7890'
if validate_phone_number(phone_number):
    print(f'{phone_number} is a valid phone number.')
else:
    print(f'{phone_number} is not a valid phone number.')

email_address = 'example@example.com'
if validate_email(email_address):
    print(f'{email_address} is a valid email address.')
else:
    print(f'{email_address} is not a valid email address.')

zip_code = '12345'
if validate_zip_code(zip_code):
    print(f'{zip_code} is a valid zip code.')
else:
    print(f'{zip_code} is not a valid zip code.')

website = 'https://www.example.com'
if validate_website(website):
    print(f'{website} is a valid website.')
else:
    print(f'{website} is not a valid website.')
```

In this example, we have defined four different functions for phone number validation, email address validation, zip code validation, and website validation using regular expressions.

Each function uses the **re** module to create a regular expression pattern for the respective domain validation. The **compile()** method is used to compile the pattern into a regular expression object. Then, the **match()** method is used to match the pattern against the input string. The **bool()** function is used to convert the match object into a Boolean value (True or False).

Finally, we have tested each validation function by calling it with a sample input and printing the result. If the validation function returns True, the input is considered valid. Otherwise, it is considered invalid.

By using regular expressions and the **re** module in Python, we can easily perform domain validation on various inputs such as phone numbers, email addresses, zip codes, and website URLs.

Using Custom Functions

Another way to validate data based on domain knowledge is by **using custom functions**. These functions can be written to check the data for specific requirements, such as checking that a date is in the correct format or that a value falls within a certain range. These functions can be implemented in Python using the built-in functions such as **strftime() or isdigit()**

Here's an example of domain validation using custom functions in Python:

```
import pandas as pd

# create sample data
data = {'date': ['2022-01-01', '2022-01-02', '2022-01-035', '2022-01-
04', '2022-01-05'],
        'time': ['12:00', '13:30', '14:15', '15:00', '1a6:30']}
df = pd.DataFrame(data)

display(df)

# define a custom function to validate date format
def validate_date_format(date):
    try:
        pd.to_datetime(date, format='%Y-%m-%d')
        return True
    except ValueError:
        return False

# define a custom function to validate time format
```

```
def validate_time_format(time):
    try:
        pd.to_datetime(time, format='%H:%M')
        return True
    except ValueError:
        return False

# apply the custom functions to validate date and time columns
df['is_date_valid'] = df['date'].apply(validate_date_format)
df['is_time_valid'] = df['time'].apply(validate_time_format)

# print the resulting dataframe
print(df)
```

The output will look like this:

	date	time
0	2022-01-01	12:00
1	2022-01-02	13:30
2	2022-01-035	14:15
3	2022-01-04	15:00
4	2022-01-05	1a6:30

```
          date   time  is_date_valid  is_time_valid
0   2022-01-01  12:00           True           True
1   2022-01-02  13:30           True           True
2  2022-01-035  14:15          False           True
3   2022-01-04  15:00           True           True
4   2022-01-05 1a6:30           True          False
```

In this example, we have created a sample dataframe with two columns - date and time. We have defined two custom functions - **validate_date_format()** and **validate_time_format()** - which use the **pd.to_datetime()** function to check whether the input string matches the specified format for date and time, respectively.

We then apply these custom functions to the date and time columns of the dataframe using the **apply()** function, and store the resulting boolean values in two new columns - **is_date_valid** and **is_time_valid**.

Finally, we print the resulting dataframe to verify that the custom functions have correctly validated the date and time columns.

Using External Libraries

Additionally, it is also possible to use external libraries, such as **nameparser** or **pyaddress**, that are specifically designed to validate certain types of data like personal name or postal address.

To use these libraries, you need to install them first using pip like below:
pip install nameparser

Here's an example of domain validation using the **nameparser** library:

```
import nameparser

# Example names to parse
name1 = "John Doe"
name2 = "Doe, Jane"

# Parse the names
parsed_name1 = nameparser.HumanName(name1)
parsed_name2 = nameparser.HumanName(name2)

# Extract components of the name
print(parsed_name1.first)
print(parsed_name1.last)
print(parsed_name1.middle)

print(parsed_name2.first)
print(parsed_name2.last)
print(parsed_name2.middle)

if parsed_name1.first and parsed_name1.last:
    print("Name is valid")
else:
    print("Name is invalid")
```

In this example, we are using the **nameparser** library to parse two names (**John Doe** and **Doe, Jane**) and extract their individual components (first name, last name, and middle name).

The **HumanName()** function takes a name string as input and returns a **HumanName** object, which has attributes like **first**, **last**, **middle**, etc. that can be used to access the different parts of the name.

This library is useful for validating and parsing various name formats such as **John Doe**, **Doe, Jane**, **Doe, John Q. Public**, **Sir James Paul McCartney CH MBE**, etc. and can be used for tasks such as address book management, personalization, and more.

Similarly, the **pyaddress** library can be used for validating and parsing physical addresses and the **phonenumbers** library can be used for validating and parsing phone numbers. These libraries provide a convenient way to validate and parse domain-specific data without having to write custom functions or regular expressions.

4.5 TIME SERIES ANALYSIS

Time series analysis is a statistical method used to analyze and model time-based data. It is a powerful tool that can be used to understand trends, patterns and relationships in data over time.

There are several different techniques used in time series analysis, including:

Decomposition: Decomposition is a technique used to break down a time series into its individual components, such as trend, seasonal, and residual components. This allows for a better understanding of the underlying patterns in the data.

Smoothing: Smoothing is a technique used to remove noise from a time series by averaging out short-term fluctuations. This can help to reveal underlying trends or patterns in the data.

Forecasting: Forecasting is a technique used to predict future values in a time series based on past data. This can be done using a variety of methods, including moving averages, exponential smoothing, and time series modeling.

Trend Analysis: Trend analysis is a technique used to identify long-term patterns and trends in a time series. This can be done using a variety of methods, such as linear regression, polynomial regression, and time series decomposition.

Seasonality Analysis: Seasonality analysis is a technique used to identify repeating patterns in a time series, such as monthly or yearly patterns. This can be done using a variety of methods, such as time series decomposition, moving averages and seasonal decomposition of time series.

Outlier Detection: Outlier detection is a technique used to identify unusual or unexpected observations in a time series. This can be done using a variety of methods, such as the Z-score, the interquartile range, and the Modified Z-score method.

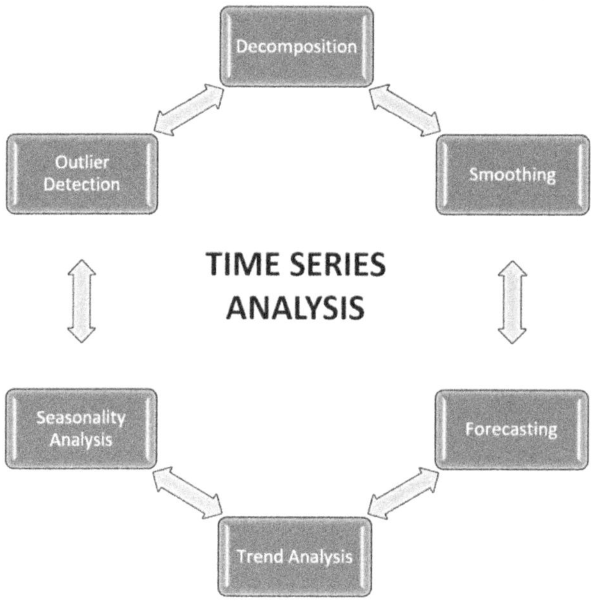

When performing time series analysis, it is important to consider the underlying assumptions of the technique being used, such as stationarity, independence, and linearity. It is also important to consider the specific characteristics of the data, such as seasonality, trend, and noise.

Time series analysis can be applied to a wide range of fields including economics, finance, engineering, weather forecasting, and many more. It is a powerful tool for understanding and predicting complex systems, and it can help organizations to make data-driven decisions.

We are not discussing this section (Time series analysis) in detail here as it needs use of some advanced library and understanding. We will discuss these in future chapters where appropriate. This section is provided for the sake of completeness.

4.6 BEST PRACTICES FOR DATA WRANGLING

Data wrangling involves transforming and preparing data for analysis. It is an essential step in the data analysis process that involves a series of steps. Here are some best practices for data wrangling that can help ensure that data is clean, consistent, and ready for analysis.

Plan ahead

- Define the problem you are trying to solve and plan your data wrangling steps accordingly.
- Identify the data sources you will use and ensure that they are reliable and complete.
- Determine the data structure you need and identify the data cleaning and transformation steps required to achieve it.

Validate data

- Check for missing or inconsistent data.
- Validate data values to ensure that they fall within the expected range.
- Check for duplicates and eliminate them if necessary.

Example: Let's say we have a dataset of customer orders that includes a column for order quantity. We should validate that the order quantity values are numeric and within a reasonable range (i.e., not negative or excessively large).

Clean data

- Remove unnecessary columns or rows that are not relevant to the analysis.
- Correct or replace missing or inconsistent data.

- Standardize data formats to ensure consistency across the dataset.

Example: In a dataset of customer addresses, we might remove columns that are not relevant to our analysis, such as the customer's phone number. We might also correct misspelled street names or replace missing ZIP codes with the correct values.

Transform data

- Convert data formats to a common standard.
- Aggregate data to the desired level of granularity.
- Create new variables that are derived from existing data.

Example: We might transform a dataset of daily sales into a monthly or quarterly summary by aggregating the data at the desired level of granularity.

Document changes

- Keep track of all changes made to the data.
- Document the reasoning behind each change.
- Store the original data and all intermediate datasets in case they are needed later.

Example: We might create a log of all data cleaning and transformation steps, including the reasoning behind each step. We should also store the original data and all intermediate datasets in case we need to refer back to them later.

Test and validate results

- Check the accuracy and completeness of the final dataset.
- Verify that the data meets the requirements for the analysis.
- Test the data using various scenarios and edge cases.

Example: We might test the final dataset by running sample analyses and verifying that the results are accurate and complete. We should also test the data using various scenarios and edge cases to ensure that it is robust and reliable.

Use version control

When working with data, it's important to keep track of changes over time. Using version control systems like Git can help you manage changes to your code and data, and make it easy to roll back to a previous version if needed.

Handle missing data appropriately

Missing data is a common issue in data wrangling. It's important to handle missing data appropriately, whether that means imputing missing values or dropping rows with missing data. Different techniques may be appropriate depending on the context and the amount of missing data.

Perform exploratory data analysis

Before diving into data wrangling, it's important to perform exploratory data analysis (EDA) to get a sense of the data and identify any issues that need to be addressed. EDA can help you identify outliers, check for skewness or other distributional issues, and explore relationships between variables.

Automate repetitive tasks

Data wrangling can involve many repetitive tasks, such as cleaning and transforming data. Automating these tasks can save time and reduce errors. Tools like Python's Pandas library and R's dplyr package can help you automate common data wrangling tasks.

By following these best practices for data wrangling, we can ensure that the data is clean, consistent, and ready for analysis.

4.7 SUMMARY

- Data wrangling is the process of cleaning, transforming, and manipulating data to make it usable for analysis.
- It is a critical step in the data analysis process, as it allows for the removal of errors and inconsistencies, and the shaping of data into a format suitable for analysis.
- Data cleaning techniques include handling missing data, removing duplicates, and detecting outliers.
- Data transformation and reshaping involve changing the structure or format of data to make it suitable for analysis.
- Data filtering and sorting can be used to select specific subsets of data for analysis.
- Data grouping and aggregation can be used to summarize and group data by certain characteristics.
- Data validation is the process of checking the data for errors and inconsistencies, and ensuring that it meets certain requirements.
- Best practices for data wrangling include an iterative process and ongoing data monitoring and maintenance.
- Hands-on exercises and real-life case studies can help to solidify understanding of data wrangling techniques and their applications.
- Data wrangling is an ongoing process, as new data is constantly being acquired and added to the dataset.
- NumPy and Pandas are widely used libraries for data wrangling in Python.
- Data wrangling is essential to ensure data quality and integrity before any analysis or modeling is performed
- It can be a time-consuming process but is crucial to the success of the overall data analysis project
- Data wrangling can be done using various tools like Python, R, SQL etc
- The ultimate goal of data wrangling is to prepare the data in a way that it can be easily analyzed and visualized
- data wrangling also includes data integration, which involves combining multiple data sources into a single dataset
- Data wrangling also includes data normalization which is the process of transforming data into a consistent format
- Data wrangling can also involve data reduction which is the process of reducing the number of data points or features in a dataset

- Data wrangling is an iterative process and requires constant monitoring, as the data may change over time.

4.8 TEST YOUR KNOWLEDGE

I. **What is the process of removing errors, inconsistencies, and irrelevant data from the data set called?**
   a) Data cleaning
   b) Data transformation
   c) Data validation
   d) Data exploration

II. **What is the process of changing the data set into a format that is suitable for the analysis called?**
   a) Data cleaning
   b) Data transformation
   c) Data validation
   d) Data exploration

III. **What is the process of checking the data for errors and inconsistencies called?**
   a) Data cleaning
   b) Data transformation
   c) Data validation
   d) Data exploration

IV. **Which Python library can be used for data exploration and visualization?**
   a) NumPy
   b) pandas
   c) Matplotlib
   d) All of the above

V. **Which Python library can be used for data cleaning?**
   a) NumPy
   b) pandas
   c) Matplotlib
   d) All of the above

VI. **Which Python library can be used for data transformation and reshaping?**

a) NumPy
b) pandas
c) Matplotlib
d) All of the above

VII. What is the process of removing duplicate data from the data set called?

a) Handling missing data
b) Removing duplicates
c) Outlier detection
d) Data cleaning

VIII. What is the process of identifying and handling outliers in the data set called?

a) Handling missing data
b) Removing duplicates
c) Outlier detection
d) Data cleaning

IX. What is the process of keeping track of the data sources, data wrangling steps, and the final data set called?

a) Data monitoring
b) Data maintenance
c) Data validation
d) Data exploration

X. What is the process of documenting the data wrangling process for reproducibility, transparency, and traceability called?

a) Data monitoring
b) Data maintenance
c) Data validation
d) Data documentation

4.9 ANSWERS

I. Answer: a) Data cleaning
II. Answer: b) Data transformation
III. Answer: c) Data validation
IV. Answer: d) All of the above

V. Answer: b) pandas
VI. Answer: b) pandas
VII. Answer: b) Removing duplicates
VIII. Answer: c) Outlier detection
IX. Answer: b) Data maintenance
X. Answer: d) Data documentation

05

5 NUMPY FOR DATA ANALYSIS

The chapter "NumPy for Data Analysis" delves into the use of NumPy, a powerful library in Python for numerical computations and data manipulation. NumPy provides a wide range of tools for working with arrays and matrices of numerical data, making it an essential tool for data analysis. In this chapter, we will explore the basics of NumPy, including its array and matrix data structures, and how to perform mathematical operations and manipulations on these structures. We will also cover advanced concepts such as broadcasting, indexing, and slicing, which allow for efficient and flexible manipulation of data. Through a series of examples and hands-on exercises, this chapter will equip readers with the knowledge and skills to effectively use NumPy for data analysis tasks.

5.1 INTRODUCTION TO NUMPY AND ITS DATA STRUCTURES

NumPy is a powerful library in Python for numerical computations and data manipulation. It provides a wide range of tools for working with arrays and matrices of numerical data, making it an essential tool for data analysis. In this section, we will explore the basics of NumPy, including its array and matrix data structures.

NumPy provides a powerful data structure called a ndarray (n-dimensional array), which is used to store and manipulate large arrays of homogeneous data (data of the same type, such as integers or floating point values). NumPy arrays are more efficient and convenient to work with compared to Python's built-in list or tuple data structures.

Creating NumPy Arrays using Initial Placeholders

In NumPy, there are several functions available for creating and initializing arrays. These functions are called initial placeholder functions or array creation functions. They allow you to create arrays of a specific shape and data type, with specified values or random values. Below are some of the commonly used initial placeholder functions in NumPy:

1. **np.zeros(shape, dtype=None, order='C'):** This function creates an array filled with zeros. The shape parameter specifies the dimensions of the array, and the dtype parameter specifies the data type of the array elements. The order parameter specifies whether the array is row-major ('C') or column-major ('F').

2. **np.ones(shape, dtype=None, order='C'**): This function creates an array filled with ones. The shape, dtype, and order parameters have the same meaning as for np.zeros().

3. **np.arange(start, stop=None, step=1, dtype=None):** This function creates an array with values in a range. The start parameter specifies the start of the range, and the stop parameter specifies the end of the range (exclusive). The step parameter specifies the step size between values. The dtype parameter specifies the data type of the array elements.

4. **np.linspace(start, stop, num=50, endpoint=True, retstep=False, dtype=None):** This function creates an array with values evenly spaced between the start and stop values. The num parameter specifies the number of values in the array, and the endpoint parameter specifies whether the stop value is included. The retstep parameter specifies whether to return the step size between values. The dtype parameter specifies the data type of the array elements.

5. **np.full(shape, fill_value, dtype=None, order='C'):** This function creates an array filled with a specified value. The shape parameter specifies the dimensions of the array, the fill_value parameter specifies the value to fill the array with, and the dtype and order parameters have the same meaning as for np.zeros().

6. **np.eye(N, M=None, k=0, dtype=<class 'float'>, order='C'):** This function creates an identity matrix with ones on the diagonal and zeros elsewhere. The N parameter specifies the number of rows, and the M parameter specifies the number of columns (if omitted, M=N). The k parameter specifies the position of the diagonal (default is 0), and the dtype and order parameters have the same meaning as for np.zeros().

7. **np.random.random(size=None):** This function creates an array with random values between 0 and 1. The size parameter specifies the dimensions of the array.

8. **np.empty(shape, dtype=float, order='C'):** This function creates an uninitialized array with random values. The shape parameter specifies the dimensions of the array, and the dtype and order parameters have the same meaning as for np.zeros().

These initial placeholder functions are very useful for creating and initializing arrays in NumPy. They provide a convenient way to create arrays of different shapes and data types, with specific values or random values.

here's a Python program that demonstrates the use of different numpy array placeholders:

```
import numpy as np

# Zeros
zeros_arr = np.zeros(5)
print("Zeros Array:")
print(zeros_arr)

# Ones
ones_arr = np.ones((2,3))
print("Ones Array:")
print(ones_arr)

# Arange
arange_arr = np.arange(5)
print("Arange Array:")
print(arange_arr)

# Linspace
linspace_arr = np.linspace(0, 1, 5)
print("Linspace Array:")
print(linspace_arr)

# Full
full_arr = np.full((2,2), 7)
print("Full Array:")
print(full_arr)

# Eye
eye_arr = np.eye(3)
print("Eye Array:")
print(eye_arr)
```

```
# Random
random_arr = np.random.random((2,3))
print("Random Array:")
print(random_arr)

# Empty
empty_arr = np.empty((2,2))
print("Empty Array:")
print(empty_arr)
```

The output will look like this:

```
Zeros Array:
[0. 0. 0. 0. 0.]
Ones Array:
[[1. 1. 1.]
 [1. 1. 1.]]
Arange Array:
[0 1 2 3 4]
Linspace Array:
[0.   0.25 0.5  0.75 1.  ]
Full Array:
[[7 7]
 [7 7]]
Eye Array:
[[1. 0. 0.]
 [0. 1. 0.]
 [0. 0. 1.]]
Random Array:
[[0.42247426 0.30512348 0.16945615]
 [0.89307503 0.59665746 0.03105797]]
Empty Array:
[[0.25 0.5 ]
 [0.75 1.  ]]
```

Creating Multidimensional Arrays

NumPy also allows us to create multidimensional arrays, such as 2-dimensional (matrices) and 3-dimensional arrays. We can create multidimensional arrays by passing a list of lists or a tuple of tuples to the array() function.

For example, here's an example program that generates arrays of various dimensions using NumPy:

```
import numpy as np

# 1D array
a1 = np.array([3, 2])

# 2D array
a2 = np.array([[1,0, 1], [3, 4, 1]])

# 3D array
a3 = np.array([[[1, 7, 9], [5, 9, 3]], [[7, 9, 9]]])

# 4D array
a4 = np.array([[[[1, 2], [3, 4]], [[5, 6], [7, 8]]], [[[9, 10], [11,
12]], [[13, 14], [15, 16]]]])

# 5D array
a5 = np.array([[[[[1, 2], [3, 4]], [[5, 6], [7, 8]]], [[[9, 10], [11,
12]], [[13, 14], [15, 16]]]],
                [[[[17, 18], [19, 20]], [[21, 22], [23, 24]]], [[[25,
26], [27, 28]], [[29, 30], [31, 32]]]]])

# Print the arrays
print("1D Array:")
print(a1)
print("2D Array:")
print(a2)
print("3D Array:")
print(a3)
print("4D Array:")
print(a4)
print("5D Array:")
print(a5)
```

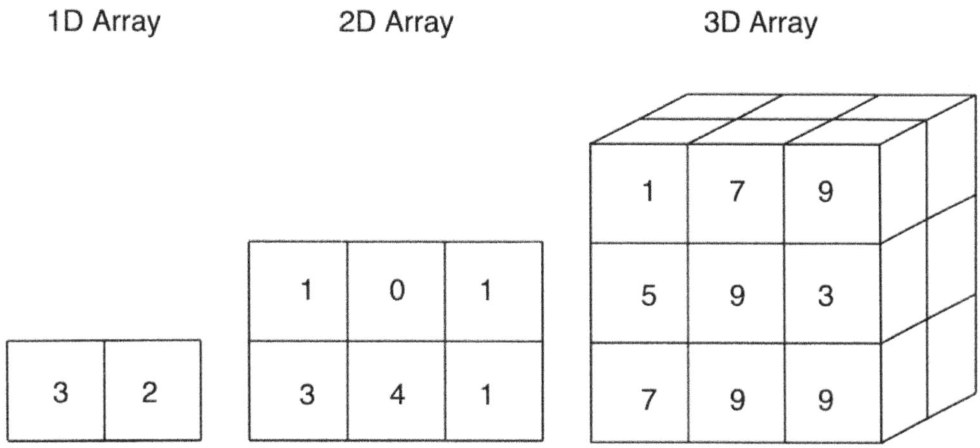

the following code creates a 2-dimensional array with 2 rows and 3 columns:

```
import numpy as np
array_2d = np.array([[1, 2, 3], [4, 5, 6]])

print(array_2d)
```

The above code will output:

[[1 2 3]

[4 5 6]]

Data Types in NumPy Arrays

NumPy is a powerful Python library used extensively for numerical computations and data analysis. It provides a rich collection of data types that can be used to represent and manipulate numerical data efficiently. In this article, we will discuss the various data types supported by NumPy and their significance.

NumPy data types are basically classes that represent a specific type of data, along with the operations that can be performed on that data. The various data types provided by NumPy can be broadly categorized into the following categories:

1. **Integer Types**: These data types are used to represent whole numbers. They can be signed or unsigned, and their size can range from 8 bits to 64 bits. Some of the commonly used integer data types in NumPy include int8, int16, int32, int64, uint8, uint16, uint32, and uint64.

2. **Floating-Point Types**: These data types are used to represent decimal numbers. They are represented in binary format and their precision and range depend on the number of bits used to represent them. Some of the commonly used floating-point data types in NumPy include float16, float32, and float64.

3. **Complex Types**: These data types are used to represent complex numbers, which have a real and imaginary part. They are represented as a pair of floating-point numbers. Some of the commonly used complex data types in NumPy include complex64 and complex128.

4. **Boolean Types**: These data types are used to represent boolean values, which can be either True or False. They are represented using a single bit, and their size is always fixed at 1 bit.

5. **String Types**: These data types are used to represent text data. They are fixed-length and can store Unicode characters. Some of the commonly used string data types in NumPy include string_, unicode_, and bytes_.

Now, let's take a look at how these data types can be used in NumPy.

Defining Data Types in NumPy:

NumPy allows us to define the data type of an array explicitly. We can do this by using the 'dtype' argument while creating the array. For example, the following code creates an array of integers:

```
import numpy as np
a = np.array([1, 2, 3, 4, 5], dtype=int)
```

Here, we have defined the data type of the array as 'int' using the 'dtype' argument.

Accessing Data Type Information:

We can access the data type information of an array using the 'dtype' attribute. For example, the following code prints the data type of the array 'a':

```
print(a.dtype)
```

Output: int32

As we can see, the 'dtype' attribute returns the data type of the array.

Converting Data Types:

NumPy also provides functions to convert the data type of an array. For example, the following code converts the data type of an array from 'int' to 'float':

```
import numpy as np
a = np.array([1, 2, 3, 4, 5], dtype=int)
b = a.astype(float)
```

Here, we have used the 'astype' function to convert the data type of the array 'a' from 'int' to 'float'.

5.2 MANIPULATING NUMPY ARRAYS

In the previous section, we introduced NumPy and its data structures, including the ndarray. In this section, we will explore the various ways to manipulate NumPy arrays, including indexing, slicing, and reshaping.

Indexing and Slicing

Indexing and slicing are powerful features in NumPy that allow you to access and manipulate specific elements or subarrays of an array. Indexing in NumPy is similar to indexing in Python lists, where you can access an element of an array using its index. For example, the following code accesses the first element of an array:

```
import numpy as np
array = np.array([1, 2, 3, 4, 5])
print(array[0])   # Output: 1
```

In addition to 1D arrays, indexing and slicing also work with multidimensional arrays. For example, the following code accesses the element at row 1, column 2 of a 2D array:

```
import numpy as np
array_2d = np.array([[1, 2, 3], [4, 5, 6], [7, 8, 9]])
print(array_2d[1, 2])   # Output: 6
```

Below, you can see some more example for indexing & slicing in the picture:

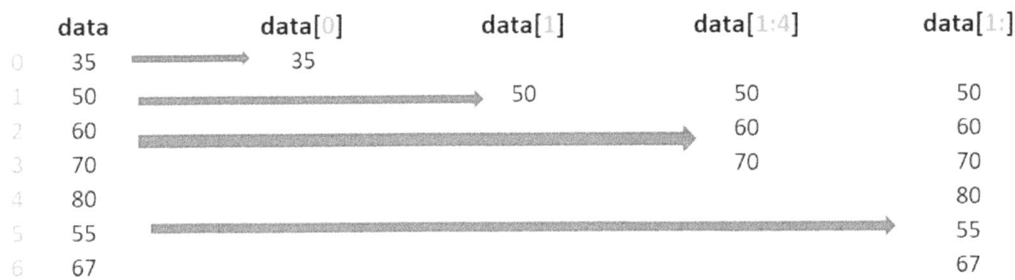

Slicing, on the other hand, allows you to access a subarray of an array. The syntax for slicing is similar to indexing, but with a colon (:) to indicate the start and end of the slice. For example, the following code creates a slice of the first three elements of an array:

```
import numpy as np
```

```
array = np.array([1, 2, 3, 4, 5])
slice = array[0:3]
print(slice)   # Output: [1 2 3]
```

Reshaping

Reshaping is the process of changing the shape or layout of an array without changing the data. NumPy provides several functions for reshaping arrays, including reshape(), ravel(), and flatten().

reshape()

The **reshape()** function allows you to change the shape of an array by specifying the number of rows and columns of the new shape. For example, the following code reshapes a 1D array with 4 elements into a 2D array with 2 rows and 2 columns:

```
import numpy as np
array = np.array([1, 2, 3, 4])
array = array.reshape(2, 2)
print(array)
```

The above code will output:

[[1 2]

[3 4]]

ravel()

The **ravel()** function returns a 1D array with all the elements of the input array. It is equivalent to reshaping the array with -1 as one of the dimensions, which tells NumPy to infer the correct size based on the other dimensions.

```
import numpy as np
array = np.array([[1, 2, 3], [4, 5, 6]])
raveled = array.ravel()
print(raveled)  # Output: [1 2 3 4 5 6]
```

flatten()

The flatten() function also returns a 1D array with all the elements of the input array, but it creates a new copy of the data, rather than returning a view of the original array.

```
import numpy as np
array = np.array([[1, 2, 3], [4, 5, 6]])
flattened = array.flatten()
print(flattened)  #[1 2 3 4 5 6]
```

The above code will output: [1 2 3 4 5 6]

In addition to reshaping and flattening arrays, NumPy also provides several other functions for manipulating arrays, such as transposing, swapping axes, and flattening them.

Transposing

The transpose() function returns a new array with the axes transposed. For example, the following code transposes a 2D array:

```
import numpy as np
array = np.array([[1, 2, 3], [4, 5, 6]])
print("Before Transpose")
print(array)
transposed = array.transpose()
print("\n After Transpose")
print(transposed)
```

The above code will output:

```
[[1 4]
 [2 5]
 [3 6]]
```

Swapping Axes:

Swapping axes in NumPy refers to the process of interchanging the rows and columns of a NumPy array. This can be achieved using the transpose() method or the swapaxes() method of NumPy. Swapping axes is an important concept in data analysis, especially when working with multi-dimensional arrays.

In NumPy, an axis refers to the dimension of a multi-dimensional array. For example, a 1-dimensional array has only one axis, while a 2-dimensional array has two axes – rows and columns. Similarly, a 3-dimensional array has three axes – height, width, and depth. The axis number starts from 0 for the first axis and increases by 1 for each additional axis.

The swapaxes() method of NumPy can also be used to swap axes of a multi-dimensional array. The swapaxes() method takes two axis numbers as arguments, which represent the axes to be swapped.

For example, to swap the axes of the NumPy array arr defined in the code below using the swapaxes() method, we can use the following code:

```
import numpy as np
arr = np.array([[1, 2], [3, 4], [5, 6]])
print("Before Swapping axes")
print(arr)

arr_swapped = arr.swapaxes(0, 1)
print("\n After Swapping axes")
print(arr_swapped)
```

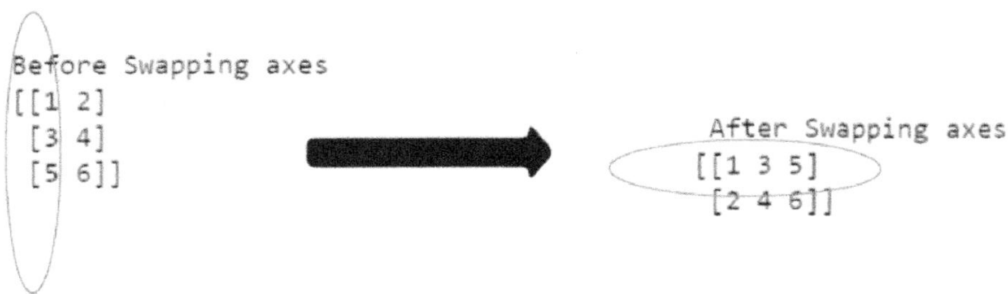

5.3 BROADCASTING

Broadcasting is important concepts in NumPy that allow for efficient and flexible manipulation of arrays. Broadcasting refers to the ability of NumPy to perform mathematical operations on arrays of different shapes, and advanced array manipulation refers to techniques for manipulating arrays in sophisticated ways.

Broadcasting is a powerful feature of NumPy that allows for mathematical operations to be performed on arrays of different shapes. It works by "stretching" or "copying" the smaller array to match the shape of the larger array, before performing the operation.

For example, the following code adds a scalar value to each element of an array:

```
import numpy as np
a = np.array([1, 2, 3])
b = 2
c = a + b
print(c)  # Output: [3 4 5]
```

In this example, the scalar value 2 is broadcasted to match the shape of the array a, before being added element-wise.

Broadcasting can also be used to perform mathematical operations between arrays of different shapes, as long as certain broadcasting rules are followed. For example, the following code multiplies a 2D array by a 1D array:

```
import numpy as np
a = np.array([[1, 2, 3], [4, 5, 6]])
b = np.array([1, 2, 3])
c = a * b
print(c)
```

It can be understood like below:

$$\begin{matrix}1 & 2 & 3\\ 4 & 5 & 6\end{matrix} \;X\; \begin{matrix}1 & 2 & 3\end{matrix} \;=\; \begin{matrix}1 & 2 & 3\\ 4 & 5 & 6\end{matrix} \;X\; \begin{matrix}1 & 2 & 3\\ 1 & 2 & 3\end{matrix} \;=\; \begin{matrix}1*1 & 2*2 & 3*3\\ 4*1 & 5*2 & 6*3\end{matrix} \;=\; \begin{matrix}1 & 4 & 9\\ 4 & 10 & 18\end{matrix}$$

In this example, the 1D array b is broadcasted to match the shape of the 2D array a, before being multiplied element-wise.

5.4 MATHEMATICAL OPERATIONS AND LINEAR ALGEBRA WITH NUMPY

NumPy is a powerful library in Python for numerical computations and data manipulation. One of its key features is its support for mathematical operations and linear algebra. In this section, we will explore the various mathematical operations and linear algebra functions that are available in NumPy, and how they can be used in data analysis tasks.

Arithmetic operations

NumPy provides a wide range of functions for performing arithmetic operations on arrays, such as addition, subtraction, multiplication, and division. These operations can be performed element-wise on arrays of the same shape, and will return an array with the same shape as the input arrays.

For example, the following code performs element-wise operation of two arrays:

```
import numpy as np
a = np.array([1, 2, 3])
b = np.array([4, 5, 6])
sum1 = a + b
print(sum1)  # Output: [5 7 9]

cross = a * b
print(cross)  # Output: [4 10 18]

subtract1 = a - b
print(subtract1)  # Output: [-3 -3 -3]

div1 = a / b
print(div1)  # Output: [0.25 0.4 0.5]
```

Linear Algebra

Linear algebra is an essential part of data science and machine learning, and Numpy offers powerful tools for working with linear algebra. In this section, we will discuss the basics of linear algebra in Numpy and provide a coding example to illustrate the concepts.

Numpy provides a set of linear algebra functions that can be used to perform various operations such as matrix multiplication, inversion, decomposition, and many more. These functions are part of the **linalg** module in Numpy and are designed to work efficiently with large arrays and matrices.

To perform linear algebra operations in Numpy, we first need to create an array or matrix. We can create an array in Numpy using the **array**() function or create a matrix using the **mat**() function. Once we have an array or matrix, we can use the various functions provided by the linalg module to perform the desired operation.

Example

Let's look at an example of performing different operation on matrix below:

```
import numpy as np

# Define matrices
A = np.array([[1, 2, 3], [4, 5, 6], [7, 8, 9]])
B = np.array([[9, 8, 7], [6, 5, 4], [3, 2, 1]])

# Matrix multiplication
C = np.dot(A, B)
print("Matrix Multiplication:")
print(C)

# Matrix inversion
D = np.linalg.inv(A)
print("\nMatrix Inversion:")
print(D)

# Eigen decomposition
E, F = np.linalg.eig(A)
print("\nEigen Decomposition:")
print("Eigenvalues:\n", E)
print("Eigenvectors:\n", F)

# Singular Value Decomposition (SVD)
G, H, I = np.linalg.svd(A)
print("\nSingular Value Decomposition:")
print("Left Singular Vectors:\n", G)
print("Singular Values:\n", H)
print("Right Singular Vectors:\n", I)

# Matrix trace
```

```
J = np.trace(A)
print("\nMatrix Trace:")
print(J)

# Determinant of a matrix
K = np.linalg.det(A)
print("\nDeterminant of a Matrix:")
print(K)

# Solving linear equations
x = np.array([[1], [2], [3]])
y = np.linalg.solve(A, x)
print("\nSolving Linear Equations:")
print(y)
```

The output will look like this:

```
Matrix Multiplication:
[[ 30  24  18]
 [ 84  69  54]
 [138 114  90]]

Matrix Inversion:
[[ 3.15251974e+15 -6.30503948e+15  3.15251974e+15]
 [-6.30503948e+15  1.26100790e+16 -6.30503948e+15]
 [ 3.15251974e+15 -6.30503948e+15  3.15251974e+15]]

Eigen Decomposition:
Eigenvalues:
 [ 1.61168440e+01 -1.11684397e+00 -3.38433605e-16]
Eigenvectors:
 [[-0.23197069 -0.78583024  0.40824829]
 [-0.52532209 -0.08675134 -0.81649658]
 [-0.8186735   0.61232756  0.40824829]]

Singular Value Decomposition:
Left Singular Vectors:
 [[-0.21483724  0.88723069  0.40824829]
 [-0.52058739  0.24964395 -0.81649658]
 [-0.82633754 -0.38794278  0.40824829]]
Singular Values:
 [1.68481034e+01 1.06836951e+00 3.33475287e-16]
Right Singular Vectors:
 [[-0.47967118 -0.57236779 -0.66506441]
 [-0.77669099 -0.07568647  0.62531805]
 [-0.40824829  0.81649658 -0.40824829]]

Matrix Trace:
15

Determinant of a Matrix:
-9.51619735392994e-16

Solving Linear Equations:
[[-0.23333333]
 [ 0.46666667]
 [ 0.1       ]]
```

This program demonstrates various operations in Linear Algebra using NumPy:

1. **Matrix multiplication**: This is demonstrated using the **np.dot()** function, which computes the dot product of two matrices. In this example, we compute the product of matrices A and B and store the result in matrix C.

2. **Matrix inversion**: This is demonstrated using the **np.linalg.inv()** function, which computes the inverse of a matrix. In this example, we compute the inverse of matrix A and store the result in matrix D.

3. **Eigen decomposition**: This is demonstrated using the **np.linalg.eig()** function, which computes the eigenvalues and eigenvectors of a matrix. In this example, we compute the eigenvalues and eigenvectors of matrix A and store the results in matrices E and F, respectively.

4. **Singular Value Decomposition (SVD)**: This is demonstrated using the **np.linalg.svd()** function, which computes the left and right singular vectors and the singular values of a matrix. In this example, we compute the SVD of matrix A and store the left singular vectors in matrix G, the singular values in matrix H, and the right singular vectors in matrix I.

5. **Matrix trace**: This is demonstrated using the **np.trace()** function, which computes the sum of the diagonal elements of a matrix. In this example, we compute the trace of matrix A and store the result in variable J.

6. **Determinant of a matrix**: This is demonstrated using the **np.linalg.det()** function, which computes the determinant of a matrix. In this example, we compute the determinant of matrix A and store the result in variable K.

7. **Solving linear equations**: This is demonstrated using the **np.linalg.solve()** function, which solves a system of linear equations of the form Ax=b. In this example, we solve the system of equations Ax=x, where x is a vector of constants, and store the result in variable y.

Eigenvalue and eigenvector

The eigenvalue and eigenvector of a matrix are important concepts in linear algebra. Eigenvalues are scalars that represent the amount by which a matrix stretches or shrinks a vector, and eigenvectors are vectors that are stretched or shrunk by a matrix.

NumPy provides the linalg.eig() function, which can be used to compute the eigenvalues and eigenvectors of a matrix.

```
import numpy as np
a = np.array([[1, 2], [3, 4]])
eigenvalues, eigenvectors = np.linalg.eig(a)
print("Eigenvalues: ", eigenvalues) # Output: Eigenvalues: [-
0.37228132 5.37228132]
print("Eigenvectors: ", eigenvectors) # Output: Eigenvectors: [[-
0.82456484 -0.41597356]
# [ 0.56576746 -0.90937671]]
```

Matrix Decomposition

Matrix decomposition is a technique for breaking a matrix down into simpler matrices. NumPy provides functions for several types of matrix decomposition, including the linalg.svd() function for singular value decomposition, and the linalg.qr() function for QR decomposition.

For example, the following code performs singular value decomposition on a matrix:

```
import numpy as np
a = np.array([[1, 2], [3, 4], [5, 6]])
U, s, V = np.linalg.svd(a)
print("U: ", U)
print("S: ", s)
print("V: ", V)
```

In addition to these examples, NumPy also provides many other mathematical operations and linear algebra functions that can be used in data analysis tasks. Understanding the capabilities of these functions and how to use them is crucial for effectively working with numerical data in Python.

As a next step, it is important to practice using these functions and concepts in real-world scenarios. One way to do this is to work through examples and tutorials that are available online. Additionally, participating in data analysis competitions, such as those on Kaggle, can provide a hands-on learning experience and help you apply your skills to real-world problems.

5.5 RANDOM SAMPLING & PROBABILITY DISTRIBUTIONS

Random sampling and probability distributions are important concepts in data analysis that allow for the generation of random numbers and the modeling of random processes. NumPy provides several functions for generating random numbers and working with probability distributions, which are useful for tasks such as statistical modeling, simulation, and machine learning.

Random Sampling

NumPy provides several functions for generating random numbers, including the random.rand() function for generating random floats in a given shape, and the random.randint() function for generating random integers in a given shape and range.

For example, the following code generates an array of 5 random floats between 0 and 1:

```
import numpy as np
np.random.rand(5)
```

and the following code generates an array of 5 random integers between 0 and 10:

```
np.random.randint(0, 10, 5)
```

NumPy also provides the random.randn() function for generating random numbers from a standard normal distribution, and the random.random_sample() function for generating random numbers from any continuous distribution.

Probability Distributions

NumPy provides functions for working with several common probability distributions, including the normal, uniform, and binomial distributions. For example, the numpy.random.normal() function can be used to generate random numbers from a normal distribution with a given mean and standard deviation, and the numpy.random.uniform() function can be used to generate random numbers from a uniform distribution within a given range.

If you don't know about the different probability distribution, here we are defining in brief for your convenience:

Normal Distribution

The normal distribution, also known as the Gaussian distribution, is a continuous probability distribution that is widely used in statistical analysis. It has a bell-shaped curve and is characterized by two parameters, the mean (μ) and the standard deviation (σ). In a normal distribution, approximately 68% of the data falls within one standard deviation of the mean, 95% falls within two standard deviations, and 99.7% falls within three standard deviations.

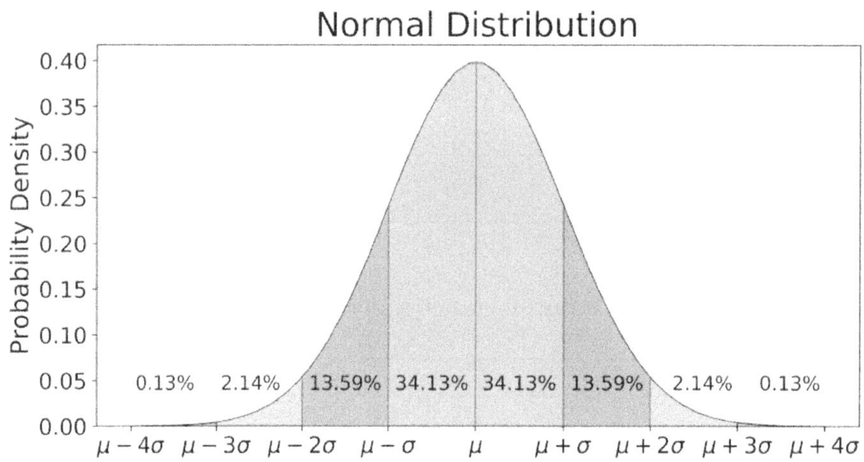

Example: Let's say we have a dataset of heights of people in a certain population. If we assume that the heights follow a normal distribution, we can find the mean and standard deviation of the dataset and plot a normal distribution curve. This curve will show us the probabilities of different height values occurring in the population.

Uniform Distribution

The uniform distribution is a probability distribution where all possible outcomes are equally likely. It is a continuous distribution where the probability of any given value occurring within a given range is proportional to the length of that range.

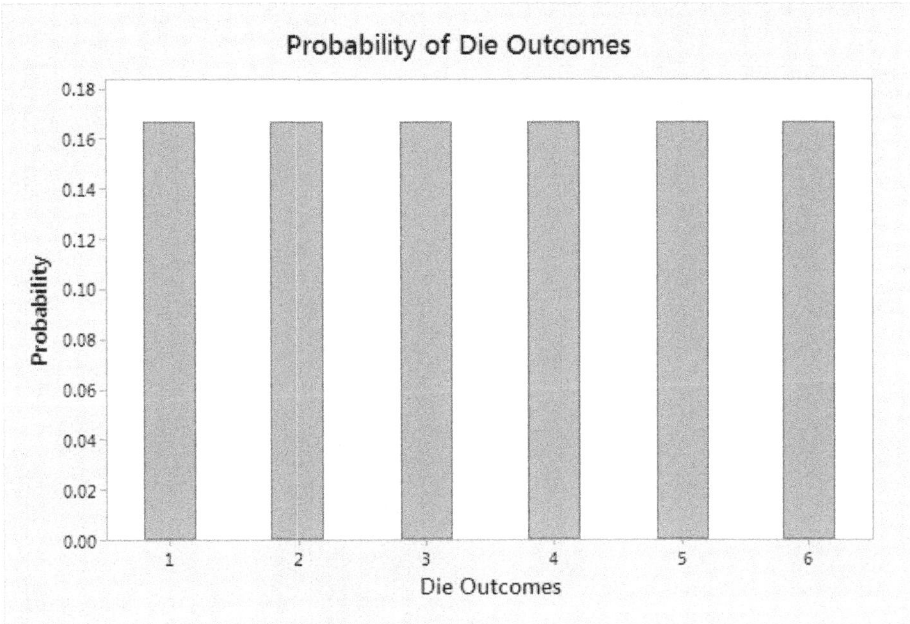

Example: Let's say we have a fair die with six faces numbered 1 to 6. If we roll the die and record the outcomes, the probability of getting any one of the six faces is the same, which is 1/6 or approximately 0.1667. This is an example of a discrete uniform distribution. In a continuous uniform distribution, the probabilities are distributed evenly over a continuous range of values.

Binomial Distribution

The binomial distribution is a probability distribution that models the number of successes in a fixed number of independent trials, where each trial has only two possible outcomes (success or failure) and the probability of success is constant across all trials. The distribution is characterized by two parameters: the number of trials (n) and the probability of success (p).

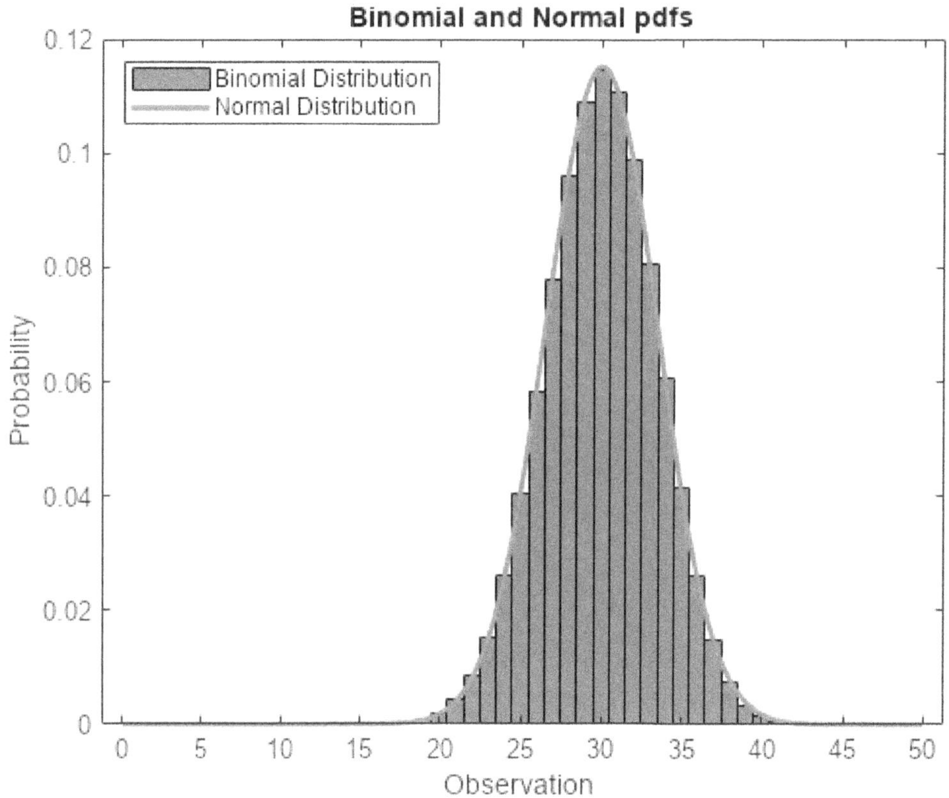

Example: Suppose we have a coin that is weighted such that the probability of getting heads is 0.6. If we flip the coin 10 times, the binomial distribution can tell us the probabilities of getting 0, 1, 2, ..., 10 heads. The probability of getting k heads in n trials is given by the formula P(k) = (n choose k) * p^k * (1-p)^(n-k), where (n choose k) is the binomial coefficient that represents the number of ways to choose k items from a set of n items.

In addition to these built-in functions, NumPy also provides the ability to create custom probability distributions using the numpy.random.generator class. This class allows you to create a generator object with a specific random number generation algorithm, and then use that generator to generate random numbers from any probability distribution.

For example, the following code creates a generator object with the Mersenne Twister algorithm, and then uses it to generate random numbers from a normal distribution:

```
gen = np.random.RandomState(seed=42)
gen.normal(loc=0, scale=1, size=5)
```

Using probability distributions and random sampling functions is important for tasks such as statistical modeling, simulation, and machine learning. For example, in statistical modeling, probability distributions are used to represent the underlying data generating process, and in simulation, random sampling is used to generate synthetic data for testing and evaluating models.

Furthermore, understanding probability distributions and their properties is important for making inferences from data. For example, if you know that the data is normally distributed, you can use the properties of the normal distribution to make inferences about the mean and standard deviation of the population. Similarly, if you know that the data is binomially distributed, you can use the properties of the binomial distribution to make inferences about the probability of success in a Bernoulli trial.

In addition, understanding probability distributions and random sampling is also important for machine learning. In supervised learning, probability distributions are often used to model the underlying data generating process, and random sampling is used to generate training and testing data. In unsupervised learning, probability distributions are used to model the underlying structure of the data, and random sampling is used to initialize the model's parameters.

Furthermore, probability distributions and random sampling are also important for data visualization. For example, histograms and probability density functions can be used to visualize probability distributions, and scatter plots and box plots can be used to visualize the relationship between variables.

5.6 USE OF NUMPY IN DATA ANALYIS

Here are some real-life use cases of NumPy in data analysis.

1. **Data Cleaning and Preprocessing**: NumPy provides functions to clean and preprocess data, such as filtering out missing values, converting data types, and scaling data. For example, if we have a dataset containing some missing values, we can use

NumPy's **nan** (not a number) value to represent those missing values and then use functions like **isnan()** and **fillna()** to handle them.

2. **Descriptive Statistics**: NumPy provides a range of functions to compute descriptive statistics on arrays, such as mean, median, mode, variance, standard deviation, and correlation coefficient. These functions are useful for exploring and summarizing data. For example, we can use NumPy's **mean()** function to compute the mean of a dataset.

3. **Data Visualization**: NumPy arrays can be used to store data for visualization purposes. NumPy arrays can be easily converted to other data structures like Pandas dataframes for visualization using libraries like Matplotlib and Seaborn. For example, we can create a NumPy array of random numbers and plot a histogram using Matplotlib.

4. **Linear Algebra**: NumPy provides a range of functions for linear algebra operations such as matrix multiplication, inversion, and decomposition. These functions are useful in analyzing data that can be represented as matrices or vectors. For example, we can use NumPy's **dot()** function to perform matrix multiplication on two arrays.

5. **Machine Learning**: NumPy is widely used in machine learning algorithms as it provides fast and efficient operations on large datasets. Machine learning algorithms often require numerical data to be represented in the form of arrays, and NumPy provides a simple and efficient way to do this. For example, we can use NumPy to create arrays of features and labels for a machine learning algorithm.

Real-life example:

Here is an example using randomly generated data:

```
import numpy as np

np.random.seed(42)

# Generate random age data for 500 people between 18 and 65
age = np.random.randint(18, 66, size=500)

# Generate random height data for 500 people between 4'6" and 7'0"
height_inches = np.random.randint(54, 84, size=500)
height_feet = height_inches / 12.0
```

```
# Generate random weight data for 500 people between 100 and 300 lbs
weight = np.random.randint(100, 301, size=500)

# Calculate BMI using the formula: weight (kg) / height (m)^2
height_meters = height_inches * 0.0254
weight_kg = weight * 0.453592
bmi = weight_kg / (height_meters ** 2)

# Print the mean, median, and standard deviation of age and BMI
print("Age:\n mean={:.2f}, \n median={:.2f},\n
std={:.2f}".format(np.mean(age), np.median(age), np.std(age)))
print("BMI: \n mean={:.2f}, \n median={:.2f}, \n
std={:.2f}".format(np.mean(bmi), np.median(bmi), np.std(bmi)))
```

The output will look like this:

```
Age:
 mean=41.98,
 median=43.00,
 std=13.79
BMI:
 mean=30.71,
 median=28.50,
 std=11.97
```

In this example, we're using numpy to generate and analyze data related to the physical characteristics of 500 people. We start by setting a random seed to ensure reproducibility of our results.

We then generate random age data for 500 people between the ages of 18 and 65, as well as random height data in inches for the same 500 people, ranging from 4'6" to 7'0". We also generate random weight data for the 500 people, ranging from 100 to 300 pounds.

Using these values, we then calculate the BMI (body mass index) for each person using the formula **weight (kg) / height (m)^2**, where weight is converted from pounds to kilograms and height is converted from inches to meters.

Finally, we use numpy to calculate the mean, median, and standard deviation of both age and BMI, which can provide insights into the overall characteristics of our dataset.

This example showcases how numpy can be used in real-life data analysis scenarios, such as health and fitness, where calculations such as BMI are commonly used to understand trends and patterns in the data.

5.7 BEST PRACTICES & PERFORMANCE TIPS FOR USING NUMPY IN DATA ANALYSIS

When working with large datasets in data analysis, it is important to use efficient and optimized techniques to manipulate and process the data. NumPy, a powerful library for numerical computing in Python, provides several best practices and performance tips for working with large datasets that can help to improve the speed and scalability of your data analysis tasks.

- **Use Vectorized Operations**: One of the most important best practices for using NumPy in data analysis is to use vectorized operations. Vectorized operations allow you to perform mathematical operations on entire arrays, rather than iterating over the elements of the array. This can greatly improve the performance of your code by taking advantage of the optimized performance of the underlying C code.
- **Avoid using for loops**: Another best practice is to avoid using for loops when working with large datasets. For loops can be slow and memory-intensive when working with large arrays, and can greatly slow down the performance of your code. Instead, use vectorized operations and built-in NumPy functions that are optimized for performance.
- **Use Broadcasting**: Broadcasting is a powerful feature of NumPy that allows you to perform mathematical operations on arrays of different shapes. Broadcasting can greatly simplify your code and improve performance by avoiding the need to reshape or tile arrays.
- **Use numpy.ndarray instead of python list**: numpy.ndarray is more efficient than python list because it stores homogeneous data in a contiguous block of memory and the element of ndarray can be accessed much faster than the elements of list.
- **Use np.array instead of np.asarray:** np.array creates a new copy of the input data, while np.asarray returns a view of the input data if possible. This can lead to unexpected behavior when modifying the input data, and can also use more memory.

- **Use np.where() instead of if-else statement**: np.where() is a more efficient way of working with conditional statements in NumPy. It can also be used as an alternative to indexing with Boolean arrays.
- **Use numpy.memmap for large arrays**: numpy.memmap can be used to handle large arrays that do not fit in memory by loading only the necessary parts of the array into memory. This can greatly improve the performance and scalability of your data analysis tasks.
- **Use np.save() and np.load() to save and load large arrays**: These functions can be used to save and load large arrays to and from disk, which can greatly improve the performance and scalability of your data analysis tasks.

5.8 SUMMARY

- NumPy is a powerful library for data analysis and scientific computing in Python
- It provides an efficient implementation of multi-dimensional arrays, which are called ndarrays
- NumPy allows for vectorized mathematical operations on arrays, making it more efficient than using loops
- NumPy provides a wide range of mathematical and statistical functions such as mean, standard deviation, correlation, and linear algebra operations
- NumPy also provides functionality for random sampling and probability distributions
- NumPy can handle large arrays efficiently, making it useful for working with large datasets
- NumPy is often used in conjunction with other data analysis libraries such as Pandas and Matplotlib
- Best practices and performance tips can be used to optimize NumPy code
- Hands-on exercises can help to solidify understanding of NumPy and its capabilities
- NumPy has been used in real-life data analysis projects such as weather data analysis.
- NumPy also provides functionality for advanced array manipulation such as broadcasting, reshaping, sorting, and filtering
- NumPy provides useful tools for data cleaning such as handling missing data, removing duplicates and detecting outliers
- NumPy provides functionality for data transformation and reshaping, allowing for easy manipulation of data into desired formats

- NumPy provides functionality for data validation, including data type validation and domain validation
- NumPy is widely used in various fields such as scientific computing, machine learning and data analysis
- NumPy is a crucial library for data scientists and analysts for its wide range of functionality and efficiency.

5.9 TEST YOUR KNOWLEDGE

I. **What is the main data structure in NumPy?**

a) Lists
b) Tuples
c) ndarrays
d) Dictionaries

II. **What is the main advantage of using NumPy over using loops for mathematical operations?**

a) NumPy is more readable
b) NumPy is more efficient
c) NumPy is easier to debug
d) NumPy is more versatile

III. **What is the function in NumPy to calculate the mean of an array?**

a) np.mean()
b) np.average()
c) np.sum()
d) np.median()

IV. **What is the function in NumPy to calculate the correlation coefficient between two arrays?**

a) np.corr()
b) np.corrcoef()
c) np.cov()
d) np.correlation()

V. **What is the function in NumPy to perform a Fast Fourier Transform on an array?**

a) np.fft()

b) np.dft()
c) np.fourier()
d) np.fft.fft()

VI. **What is the function in NumPy to generate random samples from an array?**
a) np.random.sample()
b) np.random.choice()
c) np.random.pick()
d) np.random.select()

VII. **What is the function in NumPy to fit a polynomial to an array?**
a) np.polyfit()
b) np.polyval()
c) np.fitpoly()
d) np.polynomial()

VIII. **What is the function in NumPy to save an array to disk?**
a) np.save()
b) np.dump()
c) np.export()
d) np.write()

IX. **What is the function in NumPy to reshape an array?**
a) np.reshape()
b) np.resize()
c) np.shape()
d) np.form()

X. **What is the function in NumPy to filter elements in an array based on a condition?**
a) np.filter()
b) np.where()
c) np.select()
d) np.compress()

5.10 ANSWERS

I. Answer: c) ndarrays
II. Answer: b) NumPy is more efficient

III. Answer: a) np.mean()
IV. Answer: b) np.corrcoef()
V. Answer: d) np.fft.fft()
VI. Answer: b) np.random.choice()
VII. Answer: a) np.polyfit()
VIII. Answer: a) np.save()
IX. Answer: a) np.reshape()
X. Answer: b) np.where()

06

6 PANDAS FOR DATA ANALYSIS

Welcome to the chapter on "Pandas for Data Analysis"! In this chapter, we will explore the powerful Python library Pandas, which is a fundamental tool for data manipulation and analysis. Pandas is built on top of the NumPy library and provides a high-level interface for working with data in a flexible and efficient way. With Pandas, you can easily manipulate and analyze large datasets, perform complex data transformations, and create meaningful visualizations (combining with matplotlib). This chapter will provide a comprehensive introduction to the Pandas library, including its data structures, basic operations, and advanced features. By the end of this chapter, you will have a solid understanding of how to use Pandas for data analysis tasks and be able to apply these skills in your own projects.

6.1 INTRODUCTION TO PANDAS AND ITS DATA STRUCTURES

Pandas is a powerful and widely-used open-source data analysis and manipulation library for Python. It provides high-performance, easy-to-use data structures and data analysis tools for handling and manipulating numerical tables and time series data. The two primary data structures in Pandas are the Series and DataFrame.

Series

A Series is a one-dimensional labeled array that can hold any data type. It is similar to a column in a spreadsheet or a dataset in R/Python. Each element in a Series has a unique label, called the index, which can be used to retrieve and manipulate the data. A Series can be created from a list, array, or even a scalar value.

Creating a Pandas Series:

To create a Pandas Series, we can pass a list or a NumPy array as an argument to the Series() function. Let's take an example to understand it better:

```python
import pandas as pd

# Create a Pandas Series
s = pd.Series([1, 3, 5, 7, 9])

# Print the Series
print(s)

```

```
0    1
1    3
2    5
3    7
4    9
dtype: int64
```

In the above example, we have created a Pandas Series using a list of integers. The output shows the values of the Series along with their associated index. The index starts from 0 by default, but we can also specify our own index using the index parameter.

Accessing Elements in a Series:

We can access the elements in a Pandas Series using their index. We can use either the index label or the index position to access the element. Let's take an example:

```python
import pandas as pd

# Create a Pandas Series
s = pd.Series([1, 3, 5, 7, 9], index=['a', 'b', 'c', 'd', 'e'])

# Access an element using the index label
print(s['c'])

# Access an element using the index position
print(s[3])

```

```
5
7
```

In the above example, we have created a Pandas Series with our own index labels. We have then accessed the elements using both the index label and the index position.

Operations on a Series:

We can perform various operations on a Pandas Series, such as arithmetic operations, conditional operations, and statistical operations. Let's take an example:

```
import pandas as pd

# Create a Pandas Series
s = pd.Series([1, 3, 5, 7, 9])

# Add 2 to each element in the Series
s = s + 2

# Print the updated Series
print(s)

# Get the mean of the Series
print(s.mean())

# Check if any element in the Series is greater than 8
print(s > 8)
```

```
0     3
1     5
2     7
3     9
4    11
dtype: int64
7.0
0    False
1    False
2    False
3     True
4     True
dtype: bool
```

In the above example, we have added 2 to each element in the Pandas Series using the arithmetic operation. We have then calculated the mean of the Series using the mean() function. Finally, we have checked if any element in the Series is greater than 8 using the conditional operation.

DataFrame

A DataFrame is a two-dimensional table of data with rows and columns. It is similar to a spreadsheet or a SQL table. Each column in a DataFrame is a Series and has a unique label. DataFrames can be created from a variety of sources, including lists, arrays, dicts, and even another DataFrame.

Creating a Pandas DataFrame:

A Pandas DataFrame can be created from a dictionary, a list of dictionaries, a NumPy array, or another DataFrame. Let's create a Pandas DataFrame from a dictionary:

```
import pandas as pd

data = {'name': ['John', 'Peter', 'Sara', 'David'],
        'age': [25, 30, 21, 35],
        'country': ['USA', 'Canada', 'France', 'UK']}

df = pd.DataFrame(data)
display(df)
```

	name	age	country
0	John	25	USA
1	Peter	30	Canada
2	Sara	21	France
3	David	35	UK

In this example, we created a dictionary with three keys 'name', 'age', and 'country', and their corresponding values as lists. We then passed this dictionary to the **pd.DataFrame()** function to create a Pandas DataFrame.

One of the key features of Pandas is its ability to handle missing data. It uses the special value NaN (Not a Number) to represent missing data. Pandas provide various methods for handling missing data, including fillna() and dropna().

Pandas also provides powerful data manipulation and transformation capabilities. It allows for easy reshaping, pivoting, and merging of data. It also supports powerful groupby operations, which allow you to split data into groups based on one or more columns and perform aggregate functions.

In addition, Pandas provides a wide range of data visualization capabilities through integration with popular visualization libraries such as Matplotlib and Seaborn. This makes it easy to create meaningful visualizations of your data.

6.2 READING & WRITING TO FILES USING PANDAS

In data analysis, it is often necessary to read and write data from and to various file formats. Pandas provides several functions for reading and writing data to and from different file formats, including CSV, Excel, JSON, and HDF5.

To run the code below, you should have the input file (e.g. data.csv, data.json etc.) in the same folder as your code or you should provide the absolute path of input file in the code, otherwise it will show error.

pd.read_csv(): This function is used to read a CSV file and return a DataFrame. You can specify the file path, as well as several other parameters such as the delimiter, the names of the columns, and the index column.

```
# Reading a CSV file
df = pd.read_csv('data.csv')
```

pd.read_excel(): This function is used to read an Excel file and return a DataFrame. You can specify the file path, as well as several other parameters such as the sheet name, the names of the columns, and the index column.

```
# Reading an Excel file
df = pd.read_excel('data.xlsx', sheet_name='Sheet1')
```

pd.read_json(): This function is used to read a JSON file and return a DataFrame. You can specify the file path, as well as several other parameters such as the orient of the JSON data, the names of the columns, and the index column.

```
# Reading a JSON file
df = pd.read_json('data.json', orient='columns')
```

pd.read_hdf(): This function is used to read an HDF5 file and return a DataFrame. You can specify the file path, as well as several other parameters such as the key of the data, the names of the columns, and the index column.

```
# Reading an HDF5 file
df = pd.read_hdf('data.h5', key='df')
```

df.to_csv(): This function is used to write a DataFrame to a CSV file. You can specify the file path, as well as several other parameters such as the delimiter, the names of the columns, and the index column.

```
# Writing a DataFrame to a CSV file
df.to_csv('data.csv', index=False)
```

df.to_excel(): This function is used to write a DataFrame to an Excel file. You can specify the file path, as well as several other parameters such as the sheet name, the names of the columns, and the index column.

```
# Writing a DataFrame to an Excel file
df.to_excel('data.xlsx', sheet_name='Sheet1', index=False)
```

df.to_json(): This function is used to write a DataFrame to a JSON file. You can specify the file path, as well as several other parameters such as the orient of the JSON data, the names of the columns, and the index column.

```
# Writing a DataFrame to a JSON file
df.to_json('data.json', orient='columns')
```

df.to_hdf(): This function is used to write a DataFrame to an HDF5 file. You can specify the file path, as well as several other parameters such as the key of the data, the names of the columns, and the index column.

```
# Writing a DataFrame to a Hdf file
df.to_json('data.h5, orient='columns')
```

6.3 BASIC DATAFRAME OPERATIONS

Pandas DataFrame is a two-dimensional table of data with rows and columns. It is the most widely used data structure in Pandas and is similar to a spreadsheet or a SQL table. DataFrames can be created from a variety of sources, including lists, arrays, dicts, and even another DataFrame. In this section, we will cover some basic DataFrame operations that will help you get started with using DataFrames in Pandas.

Displaying DataFrame

You can display the data in a DataFrame by simply calling the DataFrame variable. By default, the head() method shows the first 5 rows of the DataFrame, but you can also specify the number of rows you want to display. The tail() method shows the last 5 rows of the DataFrame. You can use display() or print function to show the dataframe.

```
df.head(20)
df.tail(20)
display(df)
print(df)
```

```
df.head(2)
```

	name	age	country
0	John	25	USA
1	Peter	30	Canada

```
df.tail(2)
```

	name	age	country
2	Sara	21	France
3	David	35	UK

```
display(df)
print(df)
```

	name	age	country
0	John	25	USA
1	Peter	30	Canada
2	Sara	21	France
3	David	35	UK

```
    name  age country
0   John   25     USA
1  Peter   30  Canada
2   Sara   21  France
3  David   35      UK
```

DataFrame Information

You can get information about a DataFrame by using the info() method. This method returns the number of rows and columns, the data types of each column, and the memory usage of the DataFrame.

```
df.info()
```

```
df.info()
```

```
<class 'pandas.core.frame.DataFrame'>
RangeIndex: 4 entries, 0 to 3
Data columns (total 3 columns):
 #   Column   Non-Null Count  Dtype
---  ------   --------------  -----
 0   name     4 non-null      object
 1   age      4 non-null      int64
 2   country  4 non-null      object
dtypes: int64(1), object(2)
memory usage: 224.0+ bytes
```

DataFrame Shape

You can get the shape of a DataFrame by using the shape attribute. It returns a tuple representing the number of rows and columns.

```
df.shape
```

DataFrame Columns

You can access the columns of a DataFrame by using the columns attribute. It returns an Index object containing the column labels.

```
df.columns
```

DataFrame Index

You can access the index of a DataFrame by using the index attribute. It returns an Index object containing the index labels.

```
df.index
```

DataFrame Data Types

You can check the data types of the columns in a DataFrame by using the dtypes attribute. It returns a Series with the data types of each column.

```
df.dtypes
```

DataFrame Statistics

You can get summary statistics of a DataFrame by using the describe() method. It returns a summary of the count, mean, standard deviation, minimum, and maximum of each numerical column.

```
df.describe()
```

```
1 df.describe()
```

	age
count	4.000000
mean	27.750000
std	6.075909
min	21.000000
25%	24.000000
50%	27.500000
75%	31.250000
max	35.000000

These are some of the basic DataFrame operations that you can perform in Pandas. With these basic operations, you can start working with and manipulating DataFrames in Pandas. It is a powerful tool that allows you to perform data analysis and manipulation tasks easily and efficiently.

6.4 INDEXING AND SELECTION

Indexing and selection is an essential aspect of data analysis and manipulation in Pandas. It allows you to access and manipulate specific parts of a DataFrame. Pandas provides several ways to index and select data, including:

.loc: This attribute is used to access a group of rows and columns by labels. It is primarily label based, but may also be used with a boolean array.

```
# Selecting rows by label
df.loc[1:3, ['name', 'age']]
```

.iloc: This attribute is used to access a group of rows and columns by index. It is primarily index based, but may also be used with a boolean array.

```
# Selecting rows by index
df.iloc[1:3, [0, 1]]
```

.at: This method is used to access a single value in the DataFrame by its label. It is faster than .loc for accessing a single value.

```
# Selecting a single value by label
df.at[1, 'name']
```

.iat: This method is used to access a single value in the DataFrame by its index. It is faster than .iloc for accessing a single value.

```
# Selecting a single value by index
df.iat[1, 0]
```

.ix: This attribute is used to access a group of rows and columns by either labels or index. However, it is now deprecated and it is recommended to use .loc and .iloc instead.

Boolean Indexing: This method is used to filter a DataFrame based on a boolean condition. It returns a new DataFrame containing only the rows that meet the specified condition.

```
# Filtering DataFrame based on condition
df[df['age'] > 25]
```

.query(): This method is used to filter a DataFrame based on a query expression. It is similar to Boolean indexing but it allows for more complex queries.

```
# Filtering DataFrame based on query
df.query('age > 25 and country == "UK"')
```

These are some of the methods and attributes that you can use to index and select data in Pandas. It is important to understand how to use these methods correctly, as they are essential for data analysis and manipulation tasks.

When working with large datasets, it is more memory-efficient to use the .loc and .iloc attributes instead of boolean indexing, as they return a view of the original DataFrame instead of a copy.

Indexing and selection is an important aspect of working with DataFrames in Pandas. By understanding and utilizing the various methods and attributes available for indexing and selection, you can easily access and manipulate specific parts of your data for analysis and manipulation tasks.

6.5 DATA CLEANING AND TRANSFORMATION

Data cleaning and transformation are important steps in the data analysis process. They involve cleaning and preparing the data for further analysis by removing or correcting any errors or inconsistencies. Pandas provides several methods and attributes for cleaning and transforming data, including:

.drop(): This method is used to remove rows or columns from a DataFrame. You can specify the axis (0 for rows, 1 for columns) and the labels or indexes of the rows or columns to be removed.

```
# Dropping a column
df.drop('age', axis=1)
```

.fillna(): This method is used to fill missing values in a DataFrame with a specified value or method. For example, you can use 'ffill' or 'bfill' to fill missing values with the previous or next value, respectively.

```
# Filling missing values with 0
df.fillna(0)
```

.replace(): This method is used to replace specific values in a DataFrame with a different value. You can specify the values to be replaced and the replacement value.

```
# Replacing specific values
df.replace({'USA': 'United States', 'UK': 'United Kingdom'})
```

.rename(): This method is used to rename columns or indexes in a DataFrame. You can specify a dictionary of old and new names or a function to determine the new names.

```
# Renaming columns
df.rename(columns={'name': 'full_name'})
```

.astype(): This method is used to convert the data type of a column or series. You can specify the target data type.

```
# Converting column data type
df['age'] = df['age'].astype(float)
df.head()
```

.map(): This method is used to apply a function to each element in a column or series. You can specify a function that takes one input and returns one output.

```
# Applying function to column
df['age'] = df['age'].map(lambda x: x*2)
df.head()
```

.apply(): This method is used to apply a function to each row or column in a DataFrame. You can specify a function that takes a series or DataFrame as input and returns one output.

Note that the lambda function we used here accept only numeric input and will give error for string input in the below example.

```
# Applying function to DataFrame
df.apply(lambda x: x.max() - x.min())
```

.groupby(): This method is used to group rows in a DataFrame by one or more columns. You can then apply various aggregation functions to the groups such as sum, mean, and count.

```
# Grouping and aggregating DataFrame
df.groupby('country').mean()
```

These are some of the methods and attributes that you can use to clean and transform data in Pandas. Data cleaning and transformation are important steps in the data analysis process, as they help to ensure that the data is accurate and consistent for further analysis.

For an extended list of functions, you can either search on google like "pandas functions" or you can access using the link below:

https://pandas.pydata.org/docs/reference/general_functions.html

In the future, they may change the link and it may not work, but you can always access all the functions by a simple google search.

Pandas provide several methods and attributes for cleaning and transforming data, allowing you to easily clean and prepare the data for further analysis. By understanding and utilizing the various methods and attributes available for data cleaning and transformation, you can easily remove or correct any errors or inconsistencies in your data, ensuring that the data is accurate and consistent for further analysis.

6.6 DATA EXPLORATION AND VISUALIZATION

Data exploration and visualization are important steps in the data analysis process as they allow us to understand the characteristics and patterns of our data. Pandas provides several functions and methods for data exploration and visualization, including:

df.head() and df.tail(): These functions allow us to view the first or last n rows of a DataFrame, respectively. By default, they return the first or last 5 rows, but you can specify the number of rows you want to see.

```
# Viewing the first 5 rows of a DataFrame
df.head()
# Viewing the last 3 rows of a DataFrame
```

```
df.tail(3)
```

df.info(): This function provides a summary of the DataFrame, including the number of rows and columns, the data types of each column, and the memory usage.

```
# Viewing information about a DataFrame
df.info()
```

df.describe(): This function provides summary statistics for numerical columns in the DataFrame, including the count, mean, standard deviation, minimum, and maximum.

```
# Viewing summary statistics for numerical columns
df.describe()
```

df.value_counts(): This function returns the frequency counts for each unique value in a column.

```
# Viewing the frequency counts for a column
df['column_name'].value_counts()
```

df.plot(): This function is used to create a variety of plots, including line, bar, and histogram plots, for the DataFrame. You can specify the type of plot, the x and y columns, and various other plot options.

```
import pandas as pd
import numpy as np

# Creating a pandas DataFrame with some random data
df = pd.DataFrame(np.random.randn(10, 2), columns=['Column_A',
'Column_B'])

# Creating a line plot
df.plot()
```

df.corr(): This function is used to compute pairwise correlation of columns in a DataFrame.

```
# Viewing correlation between columns
df.corr()
```

df.cov(): This function is used to compute pairwise covariance of columns in a DataFrame.

```
# Viewing covariance between columns
df.cov()
```

Keep in mind that data visualization should be used to support data exploration and not to replace it. Data visualization is an effective way to communicate insights and patterns from data, but it should always be used in conjunction with other data exploration methods to gain a comprehensive understanding of the data.

6.7 MERGING AND JOINING DATA

Merging and joining data is a common task in data analysis, and Pandas provides several functions and methods to handle it easily.

The most commonly used function for merging data in Pandas is pd.merge(). This function is used to merge or join two DataFrames on one or more columns. The basic syntax of the function is:

```
pd.merge(left_df, right_df, on='key')
```

Where left_df and right_df are the DataFrames to be merged, and on is the column or columns on which the DataFrames should be merged. By default, the function performs an inner join, which means that only the rows with matching keys in both DataFrames will be included in the merged DataFrame. However, you can also specify different types of joins such as left, right, and outer joins.

```
pd.merge(left_df, right_df, how='left', on='key')
```

Another useful function for joining data is **pd.concat().** This function is used to concatenate or join multiple DataFrames along a specific axis (rows or columns). The basic syntax of the function is:

```
pd.concat([df1, df2, df3], axis=0)
```

Where df1, df2, df3 are the DataFrames to be concatenated, and axis=0 specifies that the DataFrames should be concatenated along rows. You can also concatenate DataFrames along columns by specifying axis=1.

```
pd.concat([df1, df2, df3], axis=1)
```

Another way of joining data in Pandas is by using the join() method of DataFrames. This method is similar to the merge() function, but it can be used directly on a DataFrame, and it will join the DataFrame with another DataFrame or a Series on its index. The basic syntax of the method is:

```
df1.join(df2, on='key')
```

Where df1 is the DataFrame on which the method is called, and df2 is the DataFrame or Series to be joined.

Here's an example of merging and joining in pandas using randomly generated data:

```
import pandas as pd
import numpy as np

# Set seed for reproducibility
np.random.seed(123)

# Create two dataframes
df1 = pd.DataFrame({'Employee': ['Alice', 'Bob', 'Charlie', 'David'],
                    'Department': ['Accounting', 'Marketing',
'Engineering', 'Sales']})
df2 = pd.DataFrame({'Employee': ['Charlie', 'David', 'Edward',
'Fred'],
                    'Salary': [50000, 60000, 55000, 65000]})

display(df1)
display(df2)

# Merge dataframes based on common column 'Employee'
merged_df = pd.merge(df1, df2, on='Employee')

# Concatenate dataframes
concat_df = pd.concat([df1, df2], axis=1)
```

```
# Join dataframes based on common column 'Employee'
join_df = df1.join(df2.set_index('Employee'), on='Employee')

# Display the dataframes
print("Merged Dataframe:\n", merged_df)
print("\nConcatenated Dataframe:\n", concat_df)
print("\nJoined Dataframe:\n", join_df)
```

In the above code, we first import pandas and numpy libraries. Then, we set a seed for reproducibility. After that, we create two dataframes **df1** and **df2** with columns **Employee**, **Department**, **Salary**.

Next, we merge the two dataframes based on the common column **Employee** using the **pd.merge()** function, which returns a new dataframe containing all the rows from both dataframes where the **Employee** column matches.

We then concatenate the two dataframes using the **pd.concat()** function along the column axis (**axis=1**) to create a new dataframe with all columns from both dataframes.

Finally, we join the two dataframes based on the common column **Employee** using the **pd.DataFrame.join()** method, which returns a new dataframe containing all the rows from the left dataframe (**df1**) and the corresponding columns from the right dataframe (**df2**).

In addition to these functions and methods, Pandas also provides several options for handling missing data during joins and merges. These include the left_on, right_on, left_index, right_index, suffixes, and indicator parameters, which can be used to specify how to handle missing data, and how to handle duplicate column names in the merged DataFrame.

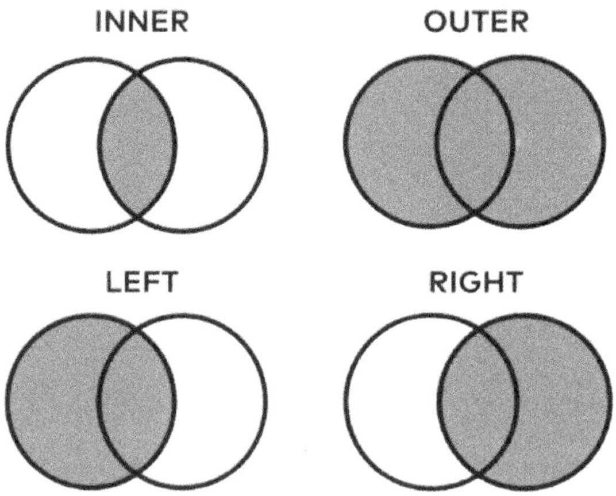

These are some of the most commonly used functions and methods for merging and joining data in Pandas. With these tools, you can easily combine and join multiple DataFrames and handle missing data in your data analysis projects.

6.8 DATA AGGREGATION WITH PANDAS

Data aggregation is the process of grouping and summarizing data based on one or more columns. Pandas provides several functions and methods for data aggregation that allows you to easily group and summarize data.

The most commonly used function for data aggregation in Pandas is groupby(). This function is used to group data based on one or more columns and perform a specific operation on the groups. The basic syntax of the function is:

```
df.groupby('column_name').operation()
```

Where df is the DataFrame, column_name is the column on which the data should be grouped, and operation is the operation to be performed on the groups, such as sum(), mean(), count(), etc.

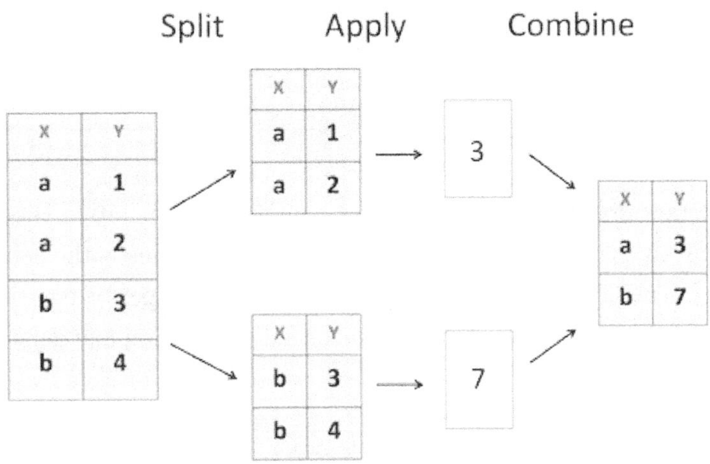

For example, to group data by 'Year' column and find the sum of 'Sales' column, we can use the following code:

```
df.groupby('Year')['Sales'].sum()
```

Another useful function for data aggregation is agg(). This function is similar to groupby() but it allows you to perform multiple operations on the groups. The basic syntax of the function is:

```
df.groupby('column_name').agg({'column1':'operation1',
'column2':'operation2'})
```

Where column1, column2 are the columns on which the operation should be performed and operation1, operation2 are the operation to be performed on the columns. For example, to group data by 'Year' column and find the mean of 'Sales' and the sum of 'Profit' columns, we can use the following code:

```
df.groupby('Year').agg({'Sales':'mean', 'Profit':'sum'})
```

Another useful function for data aggregation is pivot_table(). This function is used to create a pivot table, which is a summary of data organized in a grid format. The basic syntax of the function is:

```
pd.pivot_table(df,
```

```
                 values='column_name',
                 index='index_column',
                 columns='column_to_group',
                 aggfunc='operation')
```

Where df is the DataFrame, values is the column on which the operation should be performed, index is the column to use as the index of the pivot table, columns is the column to use as the column of the pivot table and aggfunc is the operation to be performed on the values. For example, to create a pivot table that shows the mean sales by region and gender, we can use the following code:

```
pd.pivot_table(df, values='Sales', index='Region', columns='Gender',
aggfunc='mean')
```

These are some of the most commonly used functions and methods for data aggregation in Pandas. With these tools, you can easily group and summarize data based on one or more columns and create pivot tables for data analysis.

6.9 ADVANCED STRING MANIPULATION

Advanced string manipulation is a powerful feature of Pandas that allows you to easily extract, replace, and find string patterns in data. This feature is particularly useful when working with text data, as it allows you to quickly and easily manipulate text data to gain insights.

Here are some of the advanced string manipulation methods provided by Pandas:

str.extract(): This method allows you to extract a specific pattern from a string. For example, you can use this method to extract all email addresses from a column of text data.

str.replace(): This method allows you to replace a specific pattern in a string. For example, you can use this method to replace all occurrences of a specific word in a column of text data.

str.contains(): This method allows you to check whether a string contains a specific pattern. For example, you can use this method to check whether a column of text data contains a specific word or phrase.

str.find(): This method allows you to find the index of the first occurrence of a specific pattern in a string.

str.count(): This method allows you to count the number of occurrences of a specific pattern in a string.

str.startswith(): This method allows you to check if a string starts with a specific pattern.

str.endswith(): This method allows you to check if a string ends with a specific pattern.

str.extractall(): This method allows you to extract all occurrences of a specific pattern from a string.

str.split(): This method allows you to split a string into multiple substrings based on a specific pattern.

These are some of the advanced string manipulation methods provided by Pandas. These methods allow you to easily extract, replace, and find string patterns in data. With these tools, you can quickly and easily manipulate text data to gain insights and make data-driven decisions.

6.10 TIME SERIES ANALYSIS USING PANDAS

Time series analysis using pandas is a powerful and flexible way to analyze and model time-based data. Pandas is a powerful data manipulation and analysis library for Python, and it provides a wide range of tools for working with time series data.

One of the key features of pandas is its ability to handle and manipulate time-based data using the datetime index. This allows for easy alignment of data across different time periods, and it enables the use of advanced time series functionality such as resampling and shifting.

To perform time series analysis using pandas, it is first important to ensure that the data is in a format that can be easily manipulated. This typically means that the data should be in a DataFrame format with a datetime index. Once the data is in this format, it can be easily manipulated using a variety of pandas methods.

Some of the key techniques used in time series analysis using pandas include:

Resampling: Resampling is the process of changing the frequency of a time series. This can be done using the **resample()** method in pandas, which allows for easy manipulation of data at different time frequencies, such as daily, monthly, or yearly data.

Shifting: Shifting is the process of moving data forward or backward in time. This can be done using the **shift()** method in pandas, which allows for easy manipulation of data over different time periods.

Rolling: Rolling is a method for calculating a statistic on a window of data. This can be done using the **rolling()** method in pandas, which allows for easy calculation of moving averages, standard deviations, and other statistics.

Expanding: Expanding is a method for calculating a statistic on all data up to the current point in time. This can be done using the expanding() method in pandas, which allows for easy calculation of cumulative statistics.

Time Series decomposition: Time series decomposition is the process of breaking down a time series into its individual components, such as trend, seasonal, and residual components. This can be done using the seasonal_decompose() method in pandas, which allows for easy decomposition of time series data.

Outlier Detection: Outlier detection is a technique used to identify unusual or unexpected observations in a time series. This can be done using a variety of methods, such as the Z-score, the interquartile range, and the Modified Z-score method.

Here is an example of time series analysis using pandas:

```
import pandas as pd
import numpy as np
import matplotlib.pyplot as plt

# Set the seed for reproducibility
np.random.seed(42)

# Create a datetime index with a frequency of 1 day for the year 2022
dates = pd.date_range(start='2022-01-01', end='2022-12-31', freq='D')

# Create a pandas series of random values with the datetime index
daily_sales = pd.Series(np.random.randint(100, 1000, len(dates)),
index=dates)
```

```
# Resample the series to monthly frequency and calculate the mean
value for each month
monthly_sales_mean = daily_sales.resample('M').mean()

# Plot the monthly sales means
plt.plot(monthly_sales_mean)
plt.title('Monthly Sales Mean')
plt.xlabel('Month')
plt.ylabel('Sales Mean')
plt.show()

# Create a new series with only the first 6 months of daily sales
first_half_sales = daily_sales.loc['2022-01-01':'2022-06-30']

# Resample the first half of the year to weekly frequency and
calculate the sum value for each week
weekly_sales_sum = first_half_sales.resample('W').sum()

# Plot the weekly sales sums
plt.plot(weekly_sales_sum)
plt.title('Weekly Sales Sum (Jan-Jun)')
plt.xlabel('Week')
plt.ylabel('Sales Sum')
plt.show()

# Create a new series with only the second half of the year of daily
sales
second_half_sales = daily_sales.loc['2022-07-01':'2022-12-31']

# Calculate the rolling mean with a window size of 30 days for the
second half of the year
rolling_sales_mean = second_half_sales.rolling(window=30).mean()

# Plot the rolling mean
plt.plot(rolling_sales_mean)
plt.title('Rolling Sales Mean (Jul-Dec)')
plt.xlabel('Day')
plt.ylabel('Sales Mean')
plt.show()

# Merge the monthly_sales_mean and weekly_sales_sum into a single
dataframe
sales_df = pd.concat([monthly_sales_mean, weekly_sales_sum], axis=1)

# Rename the columns for clarity
sales_df.columns = ['Monthly Sales Mean', 'Weekly Sales Sum']
```

```
# Display the merged dataframe
print(sales_df)
```

In this example, we start by generating a datetime index with a daily frequency for the year 2022 using **pd.date_range()**. We then create a pandas series of random sales values using **pd.Series()** and the datetime index. We resample the daily sales series to monthly frequency and calculate the mean value for each month using **.resample()** and **.mean()**. We then plot the resulting monthly sales means using **plt.plot()**, **plt.title()**, **plt.xlabel()**, and **plt.ylabel()**.

Next, we create a new series with only the first 6 months of daily sales using **.loc[]**. We resample this first half of the year series to weekly frequency and calculate the sum value for each week using **.resample()** and **.sum()**. We then plot the resulting weekly sales sums using **plt.plot()**, **plt.title()**, **plt.xlabel()**, and **plt.ylabel()**.

For the second half of the year, we create a new series using **.loc[]**. We calculate the rolling mean with a window size of 30 days using **.rolling()** and **.mean()**. We then plot the resulting rolling mean using **plt.plot()**, **plt.title()**, **plt.xlabel()**, and **plt.ylabel()**.

Finally, we merge the monthly sales means and weekly sales sums into a single dataframe.

By using these techniques, it is possible to perform a wide range of time series analysis using pandas. This includes identifying trends, patterns, and relationships in data over time, as well as making predictions about future data.

Before starting the analysis it's important to have a deep understanding of the dataset, and to make sure that the data is clean and in the correct format. It is also important to consider the underlying assumptions of the technique being used, such as stationarity, independence, and linearity.

6.11 BEST PRACTICES FOR USING PANDAS IN DATA ANALYSIS

Pandas is a powerful library for data analysis, but to take full advantage of its capabilities, it is important to follow best practices when using it. Here are some best practices to keep in mind when using Pandas for data analysis:

- **Keep Data Clean and Tidy:** One of the most important best practices when using Pandas is to keep data clean and tidy. This means ensuring that data is in a consistent format, that missing data is handled correctly, and that duplicate data is removed. By keeping data clean and tidy, you can ensure that your analysis is accurate and that your results are meaningful.
- **Use Vectorized Operations:** Pandas provides vectorized operations which are faster and more memory-efficient than iterating over DataFrames or Series. This means that you should use vectorized operations instead of iterating over rows and columns of data.
- **Use the inplace Parameter Wisely:** Pandas provides the inplace parameter for many data manipulation methods, which allows you to modify the original DataFrame or Series without creating a copy. However, be careful when using this parameter, as it can lead to unexpected results if you are not careful.
- **Avoid Chained Indexing:** Chained indexing occurs when you use multiple square brackets to index into a DataFrame or Series. This can cause unexpected results and is generally considered bad practice. Instead, use the .loc[], .iloc[] or .at[] methods to access data in a DataFrame or Series.
- **Use the groupby() Method Effectively:** The groupby() method is a powerful feature of Pandas that allows you to group data by one or more columns. When using this method, be sure to group by the correct columns and to use appropriate aggregation functions.
- **Use the pivot_table() Method for Cross-Tabulation:** The pivot_table() method is a powerful feature of Pandas that allows you to create cross-tabulations of data. When using this method, be sure to choose the correct values, index and columns.
- **Be Mindful of Memory Usage:** When working with large datasets, it is important to be mindful of memory usage. Try to use only the data you need, and avoid using unnecessary memory by using the appropriate data types and optimizing your code.

By following these best practices, you can ensure that you are using Pandas effectively and efficiently in your data analysis projects. This will help you to make data-driven decisions and to gain insights from your data more effectively.

6.12 SUMMARY

- Pandas is a powerful data manipulation and analysis library for Python
- Pandas provides a wide range of tools for working with time series data
- One of the key features of pandas is its ability to handle and manipulate time-based data using the datetime index
- To perform time series analysis using pandas, it is important to ensure that the data is in a format that can be easily manipulated, typically in a DataFrame format with a datetime index
- Some of the key techniques used in time series analysis using pandas include resampling, shifting, rolling, expanding, time series decomposition, and outlier detection
- These techniques can be used to identify trends, patterns, and relationships in data over time, as well as make predictions about future data
- It's important to have a deep understanding of the dataset, make sure the data is clean, and to consider the underlying assumptions of the technique being used, such as stationarity, independence, and linearity
- Time series analysis using pandas is a powerful and flexible way to analyze and model time-based data and can help organizations make data-driven decisions.
- In addition to time series analysis, Pandas also provides a wide range of functionality for data cleaning and transformation, data exploration and visualization, and data aggregation and summarization.
- Some of the key features of Pandas include its ability to handle missing data, its support for a wide range of data formats, and its powerful data indexing and selection capabilities
- Pandas also provides a wide range of functions for performing mathematical and statistical operations on data, including support for linear algebra and probability distributions
- Best practices for using Pandas in data analysis include understanding the data, cleaning and transforming it as needed, and being mindful of performance considerations when working with large data sets.

6.13 TEST YOUR KNOWLEDGE

I. What is the primary purpose of the Pandas library in Python?

a) Natural Language Processing

b) Machine Learning
c) Data Manipulation and Analysis
d) Computer Vision

II. **What is the most common data structure used in Pandas?**

a) List
b) Tuple
c) DataFrame
d) Dictionary

III. **What is the primary function of the .head() method in Pandas?**

a) Sorts the DataFrame
b) Shows the first 5 rows of the DataFrame
c) Shows the last 5 rows of the DataFrame
d) Returns the DataFrame's columns

IV. **How can missing data be handled in Pandas?**

a) By replacing it with 0
b) By dropping the rows or columns
c) By replacing it with the mean or median
d) All of the above

V. **What is the primary function of the .groupby() method in Pandas?**

a) Sorts the DataFrame
b) Groups the data by specified columns
c) Merges two DataFrames
d) Filters the DataFrame

VI. **What is the primary function of the .pivot_table() method in Pandas?**

a) Creates a pivot table from a DataFrame
b) Transposes a DataFrame
c) Groups the data by specified columns
d) Merges two DataFrames

VII. **What is the primary function of the .resample() method in Pandas?**

a) Resamples the time-series data
b) Groups the data by specified columns
c) Transposes a DataFrame
d) Merges two DataFrames

VIII. **What is the primary function of the .rolling() method in Pandas?**

a) Creates a rolling window over the data
b) Groups the data by specified columns
c) Transposes a DataFrame
d) Merges two DataFrames

IX. What is the primary function of the .merge() method in Pandas?

a) Merges two DataFrames based on common columns
b) Groups the data by specified columns
c) Transposes a DataFrame
d) Resamples the time-series data

X. What is the primary function of the .to_csv() method in Pandas?

a) Exports the DataFrame to a CSV file
b) Imports a CSV file into a DataFrame
c) Groups the data by specified columns
d) Transposes a DataFrame

6.14 ANSWERS

I. Answer: c) Data Manipulation and Analysis
II. Answer: c) DataFrame
III. Answer: b) Shows the first 5 rows of the DataFrame
IV. Answer: d) All of the above
V. Answer: b) Groups the data by specified columns
VI. Answer: a) Creates a pivot table from a DataFrame
VII. Answer: a) Resamples the time-series data
VIII. Answer: a) Creates a rolling window over the data
IX. Answer: a) Merges two DataFrames based on common columns
X. Answer: a) Exports the DataFrame to a CSV file

07

7 DESCRIPTIVE STATISTICS FOR DATA ANALYSIS

In this chapter, we will delve into the world of summarizing, describing, and understanding large sets of data. We will learn about various measures and techniques that will help us to understand the characteristics of the data and make meaningful insights. This chapter is crucial for anyone working in the field of data analysis as it forms the basis for more advanced statistical analysis. We will start by discussing the different types of descriptive statistics and then move on to the methods and techniques used to calculate them. By the end of this chapter, you will have a solid understanding of the fundamentals of descriptive statistics and be able to apply them to your own data analysis projects.

7.1 DESCRIPTIVE STATISTICS

Descriptive statistics is a branch of statistics that deals with the description and summarization of data. It is used to provide a quick and simple summary of the characteristics of a dataset. This includes measures of central tendency such as mean, median, and mode, as well as measures of spread such as range, variance, and standard deviation.

One of the main goals of descriptive statistics is to provide an understanding of the underlying data distribution. This is done by creating visualizations such as histograms, box plots, and scatter plots. These visualizations can provide insights into the shape, center, and spread of the data, as well as any outliers or patterns that may be present.

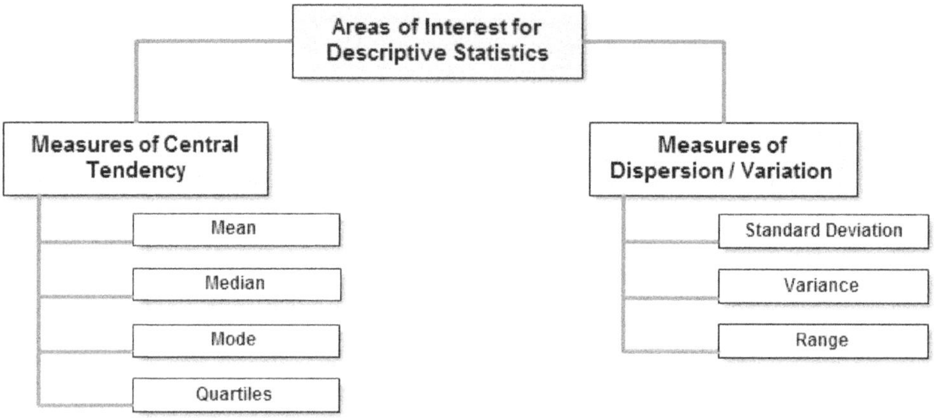

Another important aspect of descriptive statistics is the calculation of summary statistics. These statistics provide a numerical summary of the data, and include measures such as the mean, median, and mode. They can also include measures of spread such as the range, variance, and standard deviation. These summary statistics can be used to compare different datasets or to make inferences about a larger population based on a sample of data.

Descriptive statistics are an important tool in data analysis and are often used as a first step in the data analysis process. They provide a quick and simple way to understand the characteristics of a dataset, and can be used to identify patterns, trends, and outliers in the data. Additionally, they can be used to compare different datasets, or to make inferences about a larger population based on a sample of data.

7.2 MEASURES OF CENTRAL TENDENCY (MEAN, MEDIAN, MODE)

Measures of central tendency, also known as measures of central location, are used to identify the central point or typical value of a dataset. These measures include the mean, median, and mode.

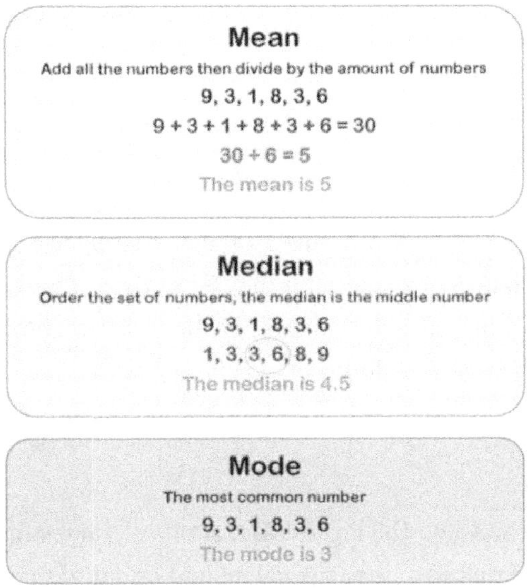

Mean

The mean, also known as the average, is the sum of all the values in a dataset divided by the number of values. It is commonly used to represent the typical value of a dataset. For example, if we have a dataset of 5 numbers: 1, 12, 3, 4, and 55, the mean would be (1+12+3+4+55)/5 = 15.

Median

The median is the middle value of a dataset when the values are arranged in numerical order. If the dataset has an odd number of values, the median is the middle value. If the dataset has an even number of values, the median is the average of the two middle values. For example, if we have a dataset of 5 numbers: 1, 2, 3, 4, and 5, the median would be 3. If we have a dataset of 6 numbers: 1, 2, 3, 4, 5, and 6, the median would be the average of 3 and 4, which is (3+4)/2 = 3.5

Mode

The mode is the value that appears most frequently in a dataset. A dataset can have one mode, multiple modes, or no mode at all. For example, if we have a dataset of 5 numbers: 1, 2, 3, 4, and 5, there is no mode because no value appears more than once. If we have a dataset of 6 numbers: 1, 2, 2, 3, 4, and 5, the mode would be 2 because it appears twice.

These measures of central tendency are not always the best way to describe a dataset and it's important to visualize the data and also use other measures such as range and standard deviation to get a better understanding of the data.

7.3 MEASURES OF SPREAD/SHAPE

Measures of spread, also known as measures of dispersion, are statistical calculations that describe the amount of variation or dispersion in a set of data. These calculations provide a way to understand how spread out the data is, and can be used to compare different sets of data. Some common measures of spread include:

Range

The range is the difference between the largest and smallest values in a set of data. It is a simple but basic measure of spread that can be easily calculated and understood.

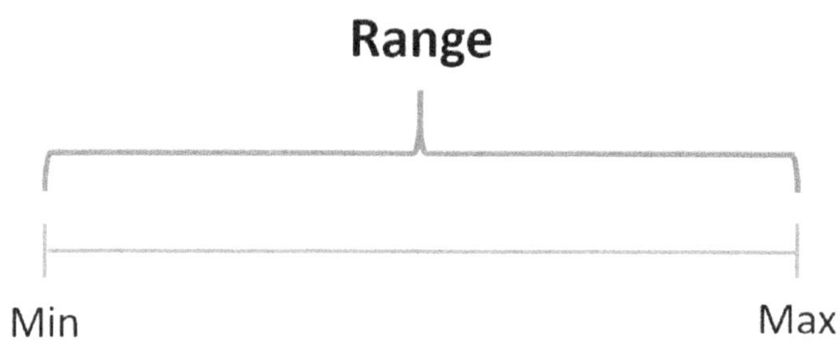

Example: In a dataset of ages (23, 25, 28, 32, 35, 42), the range would be 42-23=19.

Interquartile Range (IQR)

The IQR is a measure of spread that is based on the difference between the first and third quartiles (Q1 and Q3) of a set of data. It is a robust measure of spread that is not affected by outliers, and is commonly used in box plots.

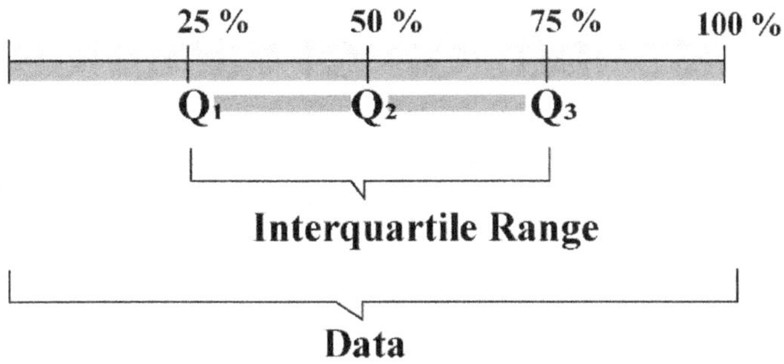

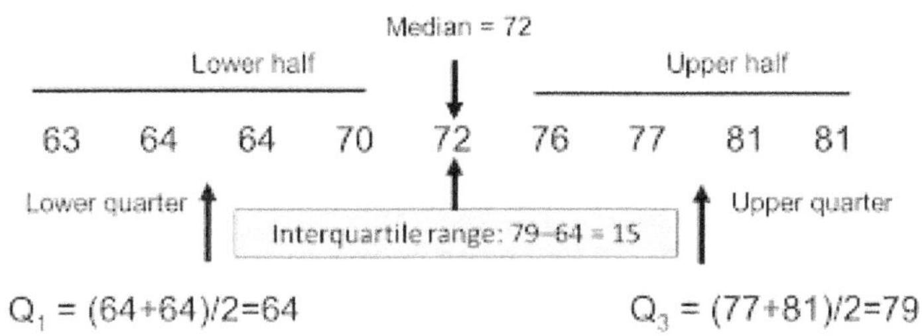

Example: In a dataset of exam scores (63, 64, 64, 70, 72, 76, 77, 81, 81), the IQR would be (79-64)=15.

Variance

The variance is a measure of spread that describes the average of the squared differences from the mean. It is a commonly used measure of spread in inferential statistics, but can be affected by outliers.

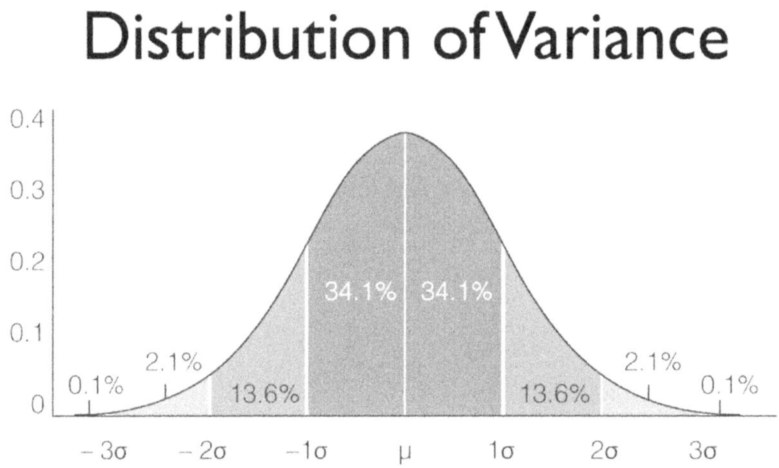

Example: In a dataset of exam scores (62, 65, 68, 70, 75, 80, 85, 90), the variance would be calculated as (((62-75)^2 + (65-75)^2 + ... + (90-75)^2)/8).

The formula for variance and standard deviation can be given as below:

$$\text{variance} = \sigma^2 = \frac{\sum (x_r - \mu)^2}{n}$$

$$\text{standard deviation} \quad \sigma = \sqrt{\frac{\sum (x_r - \mu)^2}{n}}$$

μ = mean

Standard Deviation

The standard deviation is a measure of spread that describes the average of the absolute differences from the mean. It is a commonly used measure of spread and is the square root of the variance. It is a more interpretable measure of spread as it is in the same units as the data.

Example: Using the same dataset of exam scores, the standard deviation would be the square root of the variance, which is approximately 8.5.

Percentiles

Percentiles are used to understand and describe the distribution of a dataset. They divide the data into 100 equal parts, with each part representing 1% of the data. The nth percentile is the value where n% of the data falls below it.

The formula for percentiles can be given as below:

$$P_x = \frac{x(n + 1)}{100}$$

P_x = The value at which x percentage of data lie below that value

n = Total number of observations

In respect of standard distribution of data, percentiles can be shown in bell shaped curve as below:

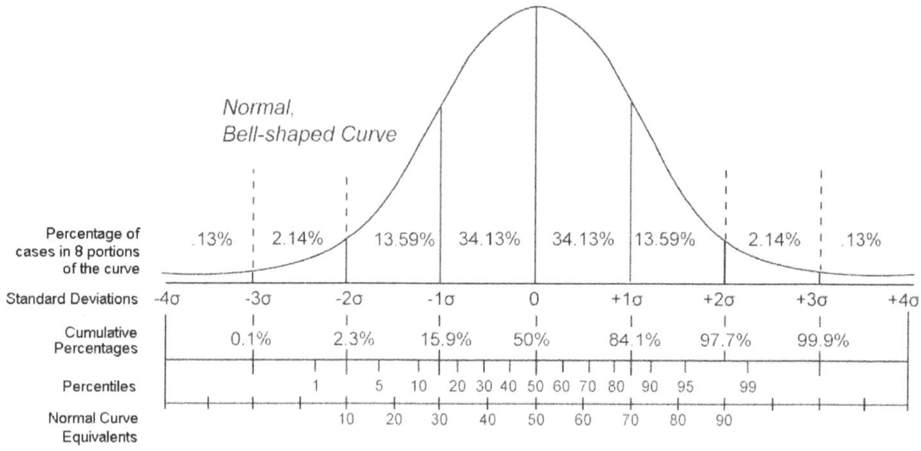

For example, consider a dataset of exam scores for a class of 20 students: [30,35,40,45,50,50,55,60,65,70,70,75,80,80,85,90,90,95,95,100]

- The 50th percentile (also known as the median) is the middle value of the dataset. In this example, it is 70, meaning that 50% of the scores fall below 70 and 50% fall above it.
- The 25th percentile is the value that separates the lowest 25% of the data from the rest. In this example, it would be 55.
- The 75th percentile is the value that separates the lowest 75% of the data from the top 25%. In this example, it would be 85.
- The 90th percentile is the value that separates the lowest 90% of the data from the top 10%. In this example, it would be 95.

Skewness

Skewness is a measure of the asymmetry of a dataset. A dataset is considered symmetric if the mean, median and mode are equal. A dataset is **positively skewed** if the mean is greater than the median, and **negatively skewed** if the mean is less than the median.

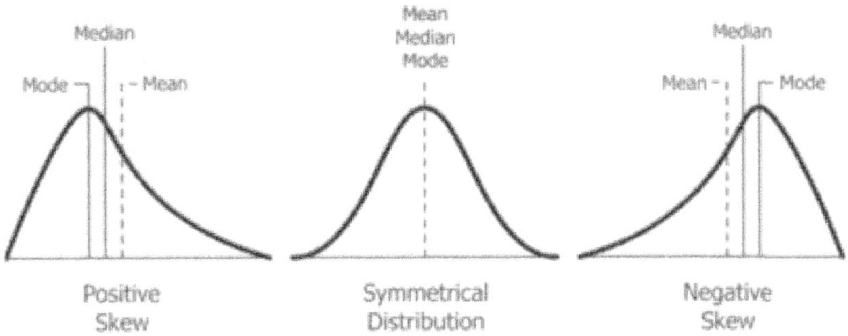

Skewness is a measure of the asymmetry of a probability distribution. A distribution is symmetric if the tail on the left side of the probability density function (pdf) is a mirror image of the tail on the right side. A distribution is skewed if the tail on one side is longer or fatter than the tail on the other side.

A positive skew means that the tail on the right side of the probability density function (pdf) is longer or fatter than the tail on the left side. This can also be referred to as a right skew. In positive skewness, the mean is greater than the median.

A negative skew means that the tail on the left side of the probability density function (pdf) is longer or fatter than the tail on the right side. This can also be referred to as a left skew. In negative skewness, the mean is less than the median.

For example, consider a dataset of ages of employees in a company. If the mean age is 35 years and the median age is 30 years, it indicates a positive skew as the mean is greater than the median. This could be because there are a few older employees with much higher ages than the rest of the employees, pulling the mean up.

On the other hand, if the mean age is 30 years and the median age is 35 years, it indicates a negative skew as the mean is less than the median. This could be because there are a few young employees with much lower ages than the rest of the employees, pulling the mean down.

Kurtosis

Kurtosis is a measure of the peakedness of a dataset. A dataset is considered normal if the kurtosis is zero. A dataset is considered **platykurtic** if the kurtosis is less than zero, and **leptokurtic** if the kurtosis is greater than zero.

Kurtosis is a measure of the "peakiness" or "flatness" of a distribution. A distribution with a high kurtosis is called leptokurtic, and has a higher peak and fatter tails, while a distribution with a low kurtosis is called platykurtic, and has a lower peak and thinner tails.

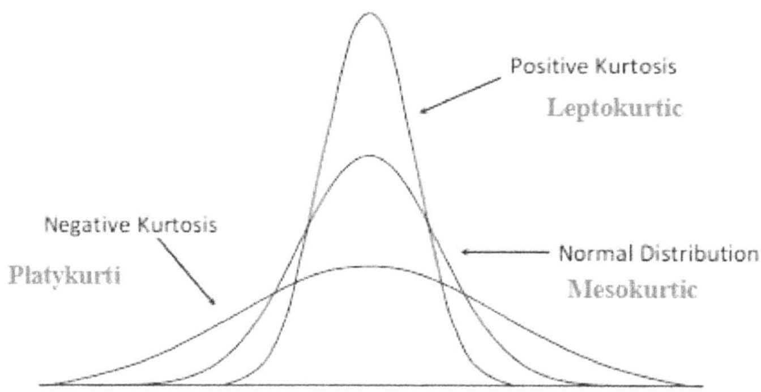

For example, consider a dataset of ages of people in a certain city. The mean age is 30, the median age is 28, and the mode age is 25. The dataset is slightly skewed to the left, as the mean is greater than the median. The kurtosis of this dataset is low, indicating that the distribution is relatively flat and does not have a high peak.

Understanding these measures of shape can help in interpreting the data and identifying patterns and trends in the data. When choosing a measure of spread, it is important to consider the characteristics of the data and the specific research question. While the range and IQR are often used in exploratory data analysis, the variance and standard deviation are more commonly used in inferential statistics. Percentiles can be used to identify and compare specific points of spread within a set of data.

7.4 FREQUENCY DISTRIBUTIONS

Frequency distributions are a way to organize and summarize a set of data by counting how many times each value appears within the dataset. This can be useful for understanding the distribution of values within the data, identifying patterns and outliers, and making comparisons between different groups or subsets of the data.

To create a frequency distribution, the first step is to identify the range of values that appear in the data. This can be done by finding the minimum and maximum values, and then dividing the range into equal intervals or "**bins.**" The number of bins used will depend on the size and complexity of the dataset, and should be chosen to provide enough detail to capture the distribution of the data while also keeping the overall picture simple and easy to understand.

Once the bins are defined, the data is sorted into the appropriate bin based on its value. The frequency of each bin is then calculated by counting the number of data points that fall into that bin. This information can then be displayed in a table or chart, such as a histogram, that shows the frequency of each bin along the y-axis and the bin values along the x-axis.

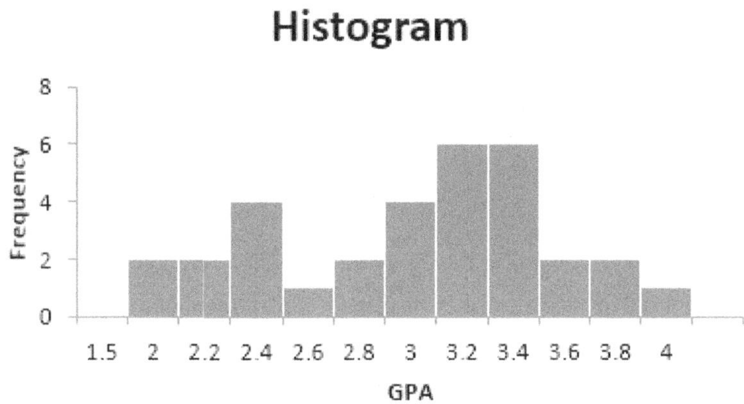

Frequency distributions can be useful for a variety of applications, such as understanding the distribution of a variable within a population, identifying patterns and outliers in the data, and making comparisons between different groups or subsets of the data. For example, a histogram of test scores for a class of students can show the distribution of scores and help identify any students who may need extra help or support. Similarly, a frequency distribution of the ages of customers at a store can help identify the store's target demographic and inform marketing and advertising strategies.

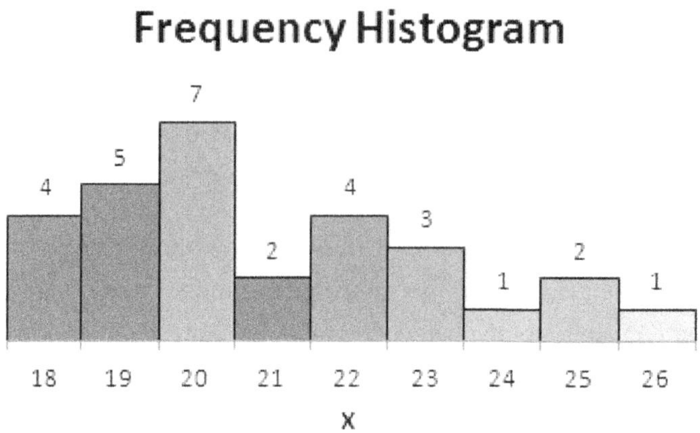

Frequency distributions are sensitive to the choice of bin size and can give different results depending on how the data is grouped. In order to avoid biases and ensure accurate results, it is essential to choose the bin size carefully and to consider the underlying distribution of the data.

One important aspect to note when creating histograms is to choose the appropriate bin size. If the bin size is too small, the histogram will appear too granular and may not provide a clear picture of the data. On the other hand, if the bin size is too large, the histogram may appear too simplistic and may not provide enough detail. It is important to strike a balance when choosing the bin size to ensure that the histogram is informative and easy to understand.

In addition, histograms are also a great tool for comparing two or more sets of data. By creating histograms for each set of data and overlaying them, it is easy to visually compare the similarities and differences between the datasets.

7.5 BOX AND WHISKER PLOTS

A Box and Whisker Plot, also known as a box plot or boxplot, is a standardized way of displaying the distribution of data based on five number summary (minimum, first quartile, median, third quartile, and maximum). It is a useful tool for comparing the distribution of data across different groups or for identifying outliers in the data.

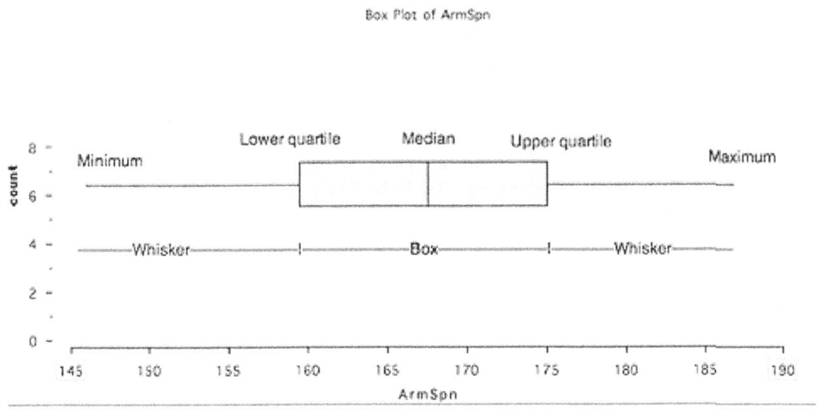

The box plot consists of a box that spans the first quartile to the third quartile (the interquartile range or IQR), a vertical line also called the "median" that indicates the median value, and two "whiskers" that extend from the box to the minimum and maximum values.

To create a box plot, the data is first divided into quartiles, or quarters of the data, using the median as a dividing point. The first quartile (Q1) is the value that separates the lowest 25% of the data from the rest, the median (Q2) is the middle value, and the third quartile (Q3) is the value that separates the highest 25% of the data from the rest.

The box in the box plot represents the interquartile range (IQR) which is the range between the first and third quartile. It is a measure of the spread of the middle 50% of the data. The box plot also includes "whiskers" which extend from the box to the minimum and maximum value of the data, indicating any outliers. Outliers are values that are more than 1.5 times the IQR away from the first or third quartile.

Box plots are commonly used in statistics and data analysis to quickly identify the **spread** and **skewness** of a data set, as well as to detect outliers. They are particularly useful when comparing multiple data sets, as the box plots can be easily superimposed for comparison.

To create a box plot in python, we can use the matplotlib library and its boxplot() function. The function takes in the data as an input and creates a box plot visualization. It is also possible to customize the appearance of the plot such as the color, line width and style of the box, whiskers and outliers. We will discuss the coding part in a future chapter.

7.6 MEASURES OF ASSOCIATION

Measures of association are statistical techniques used to determine the strength of a relationship between two variables. They are often used in descriptive statistics to analyze and understand the relationships between variables in a dataset.

There are several different measures of association, including:

- **Pearson's correlation coefficient:** This measures the linear association between two variables. It ranges from -1 to 1, with -1 indicating a perfect negative correlation, 0 indicating no correlation, and 1 indicating a perfect positive correlation.
- **Spearman's rank correlation coefficient:** This measures the monotonic association between two variables. It also ranges from -1 to 1, with the same interpretation as Pearson's coefficient.
- **Kendall's tau:** This measures the ordinal association between two variables. It ranges from -1 to 1, with the same interpretation as Pearson's and Spearman's coefficients.
- **Chi-square test:** This is a non-parametric test that is used to determine if there is a significant association between two categorical variables.
- **Point-biserial correlation coefficient:** This measures the correlation between a continuous variable and a binary variable.

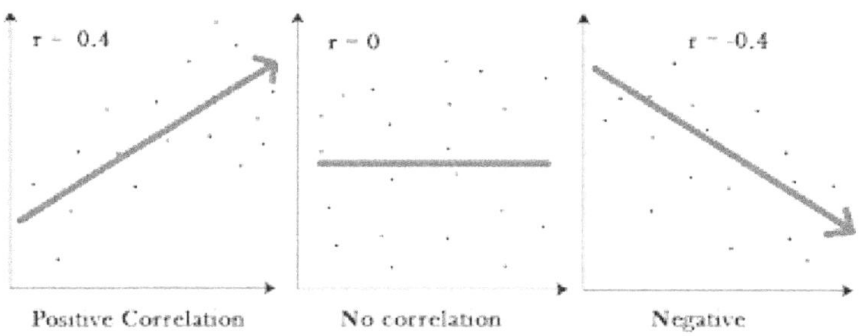

Measures of association do not indicate causality and are only used to determine the strength of a relationship between variables. Additionally, it is important to consider the assumptions and limitations of each measure when choosing which one to use for a particular dataset.

7.7 REAL-WORLD APPLICATIONS OF DESCRIPTIVE STATISTICS

Descriptive statistics is a branch of statistics that deals with the collection, presentation, and summarization of data. It plays a crucial role in data analysis and provides valuable insights into the characteristics of a dataset.

One of the key applications of descriptive statistics is in the field of business, where it is used to analyze customer data and make informed decisions. For example, a retail company may use descriptive statistics to analyze sales data and determine which products are most popular among customers. This information can then be used to make strategic decisions, such as adjusting inventory levels or developing new marketing strategies.

In the field of healthcare, descriptive statistics is used to analyze patient data and improve patient outcomes. For example, a hospital may use descriptive statistics to analyze patient outcomes after a specific treatment, in order to determine the effectiveness of the treatment.

In the field of social sciences, descriptive statistics is used to analyze survey data and understand the characteristics of a population. For example, a researcher may use descriptive statistics to analyze data from a survey on poverty, in order to understand the demographic characteristics of individuals living in poverty.

Another important application of descriptive statistics is in the field of finance, where it is used to analyze stock market data and make investment decisions. For example, an investor may use descriptive statistics to analyze the historical performance of a stock in order to determine whether it is a good investment.

Descriptive statistics is a powerful tool that is widely used in many industries and fields to analyze data and make informed decisions. It allows us to understand the characteristics of a dataset and identify patterns and trends in the data. It is a critical step in the data analysis process and can provide valuable insights for making strategic decisions and improving outcomes.

7.8 BEST PRACTICES FOR DESCRIPTIVE STATISTICAL ANALYSIS

Descriptive statistics are widely used in various fields such as business, finance, medicine, and social sciences to make sense of large amounts of data. In this section, we will discuss the best practices for descriptive statistical analysis.

- **Understand the Data**: Before performing any statistical analysis, it is important to understand the data that you are working with. This includes understanding the variables, the measurement scales, and the distribution of the data.
- **Choose the Right Measures**: Different measures of central tendency, spread, and shape are suitable for different types of data. For example, the mean is affected by outliers, so it is not a suitable measure for skewed data. Similarly, the standard deviation is not suitable for data on a categorical scale.
- **Use Graphical Methods**: Graphical methods such as histograms, box plots, and scatter plots are useful in understanding the distribution and patterns in the data. They provide a visual representation of the data and help identify outliers and skewness.
- **Use Summary Statistics**: Summary statistics such as the mean, median, and mode provide a concise summary of the data. They are useful in understanding the central tendency of the data.
- **Use Inferential Statistics**: Inferential statistics is used to make predictions about a population based on a sample. Inferential statistics can be used to test hypotheses and make predictions about the population.
- **Communicate Results**: Communicating the results of the descriptive statistical analysis is an important step. It is important to use clear and simple language when communicating the results. Use tables, graphs, and charts to make the results more understandable.
- **Be aware of outliers**: Outliers can have a significant impact on the results of the descriptive statistical analysis. It is important to identify and handle outliers appropriately.
- **Use Software**: There are many software packages available for performing descriptive statistical analysis. These software packages can save time and make the analysis more accurate.
- **Keep Learning**: Descriptive statistics is a constantly evolving field. It is important to keep learning new techniques and methods to improve your data analysis skills.

By following these best practices, you can ensure that your descriptive statistical analysis is accurate, clear, and informative.

7.9 SUMMARY

- Descriptive statistics is a branch of statistics that deals with summarizing and describing data.
- The main goal of descriptive statistics is to provide a summary of the main characteristics of the data, such as the central tendency, spread, and shape of the data.
- Measures of central tendency include mean, median, and mode, and are used to describe the "middle" of the data.
- Measures of spread include range, variance, and standard deviation, and are used to describe how spread out the data is.
- Measures of shape include skewness and kurtosis, and are used to describe the shape of the data distribution.
- Frequency distributions and histograms are tools used to visualize the data and understand the distribution of values.
- Box and whisker plots are another tool used to visualize the distribution of data, particularly the spread and skewness.
- Measures of association, such as correlation, are used to understand the relationship between two or more variables.
- Descriptive statistics can be used in many real-world applications, such as market research, finance, and healthcare.
- It is important to follow best practices when conducting descriptive statistical analysis, such as using appropriate measures for the type of data and understanding the limitations of the analysis.

7.10 TEST YOUR KNOWLEDGE

I. What is the measure of central tendency that is most affected by outliers?

a) Mean
b) Median
c) Mode
d) Range

II. What is the formula for calculating the standard deviation?

a) (1/n) * Σ(x - μ)^2
b) (1/n) * Σ(x - Σx)^2
c) (1/n) * Σx
d) Σ(x - μ)^2

III. **What type of data is best represented by a histogram?**
  a) Continuous data
  b) Categorical data
  c) Ordinal data
  d) Nominal data

IV. **What is the difference between a positive and negative correlation?**
  a) Positive correlation means that the variables increase together, while negative correlation means that one variable increases while the other decreases.
  b) Positive correlation means that the variables decrease together, while negative correlation means that one variable decreases while the other increases.
  c) Positive correlation means that the variables are unrelated, while negative correlation means that the variables are related in some way.
  d) Positive correlation means that the variables are related in some way, while negative correlation means that the variables are unrelated.

V. **What is the definition of skewness?**
  a) The degree of peakedness in a distribution
  b) The degree of asymmetry in a distribution
  c) The degree of spread in a distribution
  d) The degree of centerness in a distribution

VI. **What is the difference between a histogram and a bar chart?**
  a) Histograms are used for continuous data, while bar charts are used for categorical data.
  b) Histograms are used for categorical data, while bar charts are used for continuous data.
  c) Histograms are used for ordinal data, while bar charts are used for nominal data.
  d) Histograms are used for nominal data, while bar charts are used for ordinal data.

VII. **What is the definition of kurtosis?**
  a) The degree of peakedness in a distribution
  b) The degree of asymmetry in a distribution
  c) The degree of spread in a distribution
  d) The degree of centerness in a distribution

VIII. **What measure of association is used to determine the strength and direction of a linear relationship between two variables?**
  a) Correlation coefficient
  b) Covariance
  c) Chi-square test

d) T-test

IX. What is the difference between a box plot and a whisker plot?

a) Box plots include the median and quartiles, while whisker plots only include the minimum and maximum values.

b) Box plots include the mean and standard deviation, while whisker plots only include the minimum and maximum values.

c) Box plots include the range and interquartile range, while whisker plots only include the minimum and maximum values.

d) Box plots only include the minimum and maximum values, while whisker plots include the median and quartiles.

X. What is the formula for calculating the standard deviation?

a) (x - mean) / n

b) (x - mean)^2 / n

c) (x - mean)^2 / (n-1)

d) (x - mean) / (n-1)

XI. What is the difference between skewness and kurtosis?

a) Skewness measures the symmetry of a distribution, while kurtosis measures the peakedness of a distribution.

b) Skewness measures the peakedness of a distribution, while kurtosis measures the symmetry of a distribution.

c) Skewness measures the spread of a distribution, while kurtosis measures the center of a distribution.

d) Skewness measures the outliers in a distribution, while kurtosis measures the skewness of a distribution.

XII. What measure of spread shows the range of values within a dataset?

a) Range

b) Variance

c) Standard deviation

d) Interquartile range

XIII. Which measure of central tendency is affected by outliers in a dataset?

a) Mode

b) Median

c) Mean

d) All of the above

XIV. What is the formula for calculating skewness in a dataset?

a) (3(mean - median)) / standard deviation
b) (mean - median) / standard deviation
c) (3(mean - mode)) / standard deviation
d) (mean - mode) / standard deviation

XV. Which type of plot is best used for displaying the distribution of a dataset?

a) Line plot
b) Bar plot
c) Histogram
d) Pie chart

XVI. What is the name of the measure used to determine the strength and direction of a linear relationship between two variables?

a) Range
b) Variance
c) Correlation coefficient
d) Standard deviation

XVII. What is the range of values for the correlation coefficient?

a) -1 to 1
b) 0 to 1
c) -100 to 100
d) 1 to 100

XVIII. What is the formula for calculating the variance of a dataset?

a) (sum of (data point - mean)^2) / (number of data points)
b) (sum of (data point - median)^2) / (number of data points)
c) (sum of (data point - mode)^2) / (number of data points)
d) (sum of (data point - range)^2) / (number of data points)

XIX. What measure of association is used to determine the strength of a relationship between two categorical variables?

a) Correlation coefficient
b) Chi-squared test
c) T-test
d) ANOVA

XX. What is the name of the measure used to determine the proportion of variation in one variable that can be explained by another variable?

a) Correlation coefficient
b) Coefficient of determination
c) T-test
d) ANOVA

XXI. What is the purpose of a box and whisker plot?

a) To display the distribution of a dataset
b) To show the spread of a dataset and identify outliers
c) To determine the correlation between two variables
d) To compare the means of multiple datasets.

7.11 ANSWERS

I. Answer: A. Mean
II. Answer: A. (1/n) * Σ(x - μ)^2
III. Answer: A. Continuous data
IV. Answer: A. Positive correlation means that the variables increase together, while negative correlation means that one variable increases while the other decreases.
V. Answer: B. The degree of asymmetry in a distribution
VI. Answer: A. Histograms are used for continuous data, while bar charts are used for categorical data.
VII. Answer: A. The degree of peakedness in a distribution
VIII. Answer: A. Correlation coefficient
IX. Answer: a) Box plots include the median and quartiles, while whisker plots only include the minimum and maximum values.
X. Answer: c) (x - mean)^2 / (n-1)
XI. Answer: a) Skewness measures the symmetry of a distribution, while kurtosis measures the peakedness of a distribution.
XII. Answer: a) Range
XIII. Answer: c) Mean
XIV. Answer: a) (3(mean - median)) / standard deviation
XV. Answer: c) Histogram
XVI. Answer: c) Correlation coefficient
XVII. Answer: a) -1 to 1
XVIII. Answer: a) (sum of (data point - mean)^2) / (number of data points)

XIX. Answer: b) Chi-squared test

XX. Answer: b) Coefficient of determination

XXI. Answer: b) To show the spread of a dataset and identify outliers

08

8 DATA EXPLORATION

Data exploration is an essential step in the data analysis process. It involves understanding the characteristics of the data, identifying patterns and relationships, and uncovering insights that can inform further analysis and decision making. In this chapter, we will cover the basics of data exploration, including univariate, bivariate, and multivariate analysis, as well as various data visualization techniques. We will also look at advanced techniques for identifying patterns and relationships in data, as well as best practices for data exploration. By the end of this chapter, readers will have a solid understanding of how to approach and conduct data exploration, and be well-equipped to uncover valuable insights from their data.

8.1 INTRODUCTION TO DATA EXPLORATION

Data exploration is the process of analyzing, summarizing, and visualizing data in order to understand its characteristics and identify patterns, trends, and relationships. It is an essential step in the data analysis process and enables data scientists and analysts to gain a deeper understanding of their data before moving on to more advanced analysis or model building.

There are several key steps involved in data exploration, including:

Data preparation: This involves cleaning and transforming the data to make it ready for analysis. This can include removing missing or duplicate values, handling outliers, and converting data into the appropriate format.

Univariate analysis: This involves analyzing and summarizing individual variables or features in the data. This can include measures of central tendency, such as mean, median, and mode, as well as measures of dispersion, such as standard deviation and range.

Bivariate analysis: This involves analyzing the relationship between two variables in the data. This can include calculating the correlation coefficient, creating scatter plots, and performing statistical tests to determine if there is a significant relationship between the variables.

Multivariate analysis: This involves analyzing the relationship between three or more variables in the data. This can include techniques such as principal component analysis and cluster analysis.

Data visualization: This involves creating visual representations of the data in order to identify patterns, trends, and relationships. This can include techniques such as bar charts, histograms, scatter plots, and heat maps.

Identifying patterns and relationships: This involves analyzing the data to identify patterns and relationships that may not be immediately apparent. This can include techniques such as hypothesis testing and model building.

By following these steps, data scientists and analysts can gain a deeper understanding of their data and identify insights that can inform further analysis and decision making.

Data exploration is an iterative process, and analysts may need to go back and repeat steps as needed in order to fully understand the data.

8.2 UNIVARIATE ANALYSIS

Univariate analysis is the simplest form of data analysis and involves the analysis of single variables or features in a dataset. It is used to understand the distribution and characteristics of individual variables, and can provide valuable insights into the data.

There are several key steps involved in univariate analysis, including:

- **Descriptive statistics:** This involves calculating measures of central tendency such as mean, median, and mode, as well as measures of dispersion such as standard deviation and range. These statistics provide a basic understanding of the distribution of the data and can be used to identify outliers or skewness.

- **Frequency distribution:** This involves counting the number of occurrences of each unique value in a variable. This can be used to identify the most common values and the distribution of the data.
- **Histograms and density plots:** These are graphical representations of the frequency distribution of a variable. Histograms show the frequency distribution of a variable in bins or intervals, while density plots show the probability density function of a variable. Both provide a visual representation of the distribution of the data and can be used to identify patterns and outliers.
- **Box plots:** Box plots are a standardized way of displaying the distribution of data. They show the median, quartiles, and range of the data, as well as any outliers. Box plots can be used to quickly identify skewness and outliers in a variable.
- **Normality tests:** This involves testing if a variable is normally distributed. Normality tests can be used to check if a variable's distribution is close to a normal distribution.

By performing univariate analysis, data scientists and analysts can gain a deeper understanding of the distribution and characteristics of individual variables in the data. This can inform further analysis and decision making.

Univariate analysis should be performed on each variable in the dataset in order to gain a comprehensive understanding of the data.

8.3 BIVARIATE ANALYSIS

Bivariate analysis is the process of analyzing the relationship between two variables in a dataset. It is a step beyond univariate analysis, which only looks at individual variables, and can provide a deeper understanding of the relationship between variables and how they may be related to one another.

There are several key steps involved in bivariate analysis, including:

Scatter plots:

Scatter plots are a powerful tool for bivariate analysis, which allows us to study the relationship between two continuous variables. A scatter plot is a graphical representation of a set of data points that are plotted on two axes, with each point representing a pair of values for the two variables being studied.

Scatter plots are commonly used in data analysis to visualize the correlation between two variables. The pattern of the points on the scatter plot can give insights into the strength and direction of the relationship between the two variables. If the points on the scatter plot form a linear pattern, then the variables are said to have a linear relationship.

To illustrate the use of scatter plots for bivariate analysis, let's consider an example. Suppose we have a dataset that contains information on the height and weight of individuals. We can use a scatter plot to study the relationship between these two variables.

Here is an example of scatter plot analysis using randomly generated data:

```
import numpy as np
import pandas as pd
import matplotlib.pyplot as plt

np.random.seed(123)
height = np.random.normal(loc=170, scale=10, size=100)
weight = np.random.normal(loc=70, scale=10, size=100)
age = np.random.normal(loc=30, scale=5, size=100)

df = pd.DataFrame({'height': height, 'weight': weight, 'age': age})

# Scatter plot of height and weight
plt.scatter(df['height'], df['weight'])
plt.title('Height vs Weight')
plt.xlabel('Height')
plt.ylabel('Weight')
plt.show()

# Scatter plot of height and age
plt.scatter(df['height'], df['age'])
plt.title('Height vs Age')
plt.xlabel('Height')
plt.ylabel('Age')
plt.show()

# Scatter plot of weight and age
plt.scatter(df['weight'], df['age'])
plt.title('Weight vs Age')
plt.xlabel('Weight')
```

```
plt.ylabel('Age')
plt.show()
```

The output will look like this:

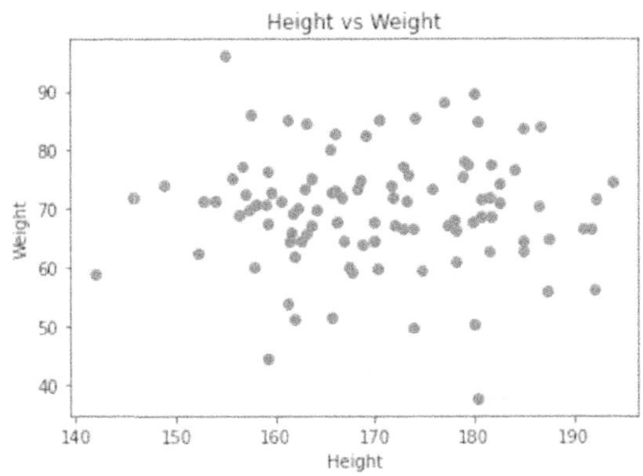

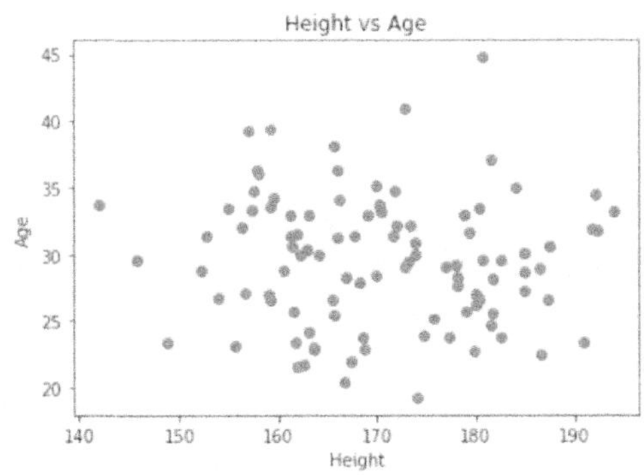

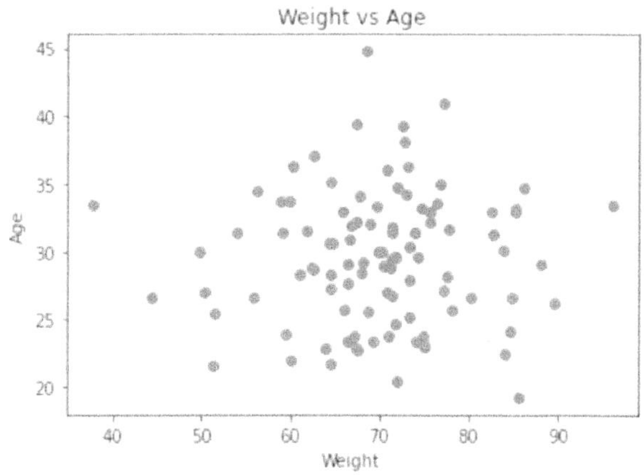

In this example, we first generate three sets of randomly generated data for height, weight, and age. We then combine these into a pandas dataframe called **df**.

Next, we create three scatter plots using **plt.scatter()**. The first scatter plot compares height and weight, the second scatter plot compares height and age, and the third scatter plot compares weight and age.

The resulting plots show the relationship between each pair of variables. The scatter plot of height and weight shows a moderate positive correlation between the two variables, while the scatter plot of height and age shows no clear relationship between the variables. The scatter plot of weight and age also shows no clear relationship between the variables.

These scatter plots are useful for identifying potential patterns and relationships between variables in a dataset. By analyzing the scatter plots, we can gain insight into the underlying data and make informed decisions about further analysis or modeling.

Correlation:

Correlation analysis is a statistical technique used to measure the strength and direction of the relationship between two variables. It helps in understanding how one variable is related to another and how changes in one variable affect the other variable. In this article, we will discuss the correlation for bivariate analysis and provide an example using randomly generated data and seed.

Bivariate analysis involves the analysis of two variables simultaneously to determine the relationship between them. The correlation coefficient is used to measure the strength and direction of the relationship between two variables. The correlation coefficient ranges from -1 to 1. A correlation coefficient of 1 indicates a perfect positive correlation, a correlation coefficient of 0 indicates no correlation, and a correlation coefficient of -1 indicates a perfect negative correlation.

Suppose we want to explore the relationship between a company's advertising budget and its sales revenue. We can collect data for these two variables over a period of time and use Python's NumPy and Pandas libraries to analyze the correlation between them.

Here is an example of using correlation for bivariate analysis:

```
import numpy as np
import pandas as pd
import matplotlib.pyplot as plt

np.random.seed(123)

ad_budget = np.random.normal(100, 20, 50)
sales_revenue = ad_budget * np.random.normal(5, 1, 50)

data = pd.DataFrame({'ad_budget': ad_budget, 'sales_revenue':
sales_revenue})
corr = data['ad_budget'].corr(data['sales_revenue'])
print("Correlation coefficient:", corr)
plt.scatter(data['ad_budget'], data['sales_revenue'])
plt.title("Relationship between Advertising Budget and Sales Revenue")
plt.xlabel("Advertising Budget")
plt.ylabel("Sales Revenue")
plt.show()
```

The output will look like this:

```
Correlation coefficient: 0.7236592382940178
```

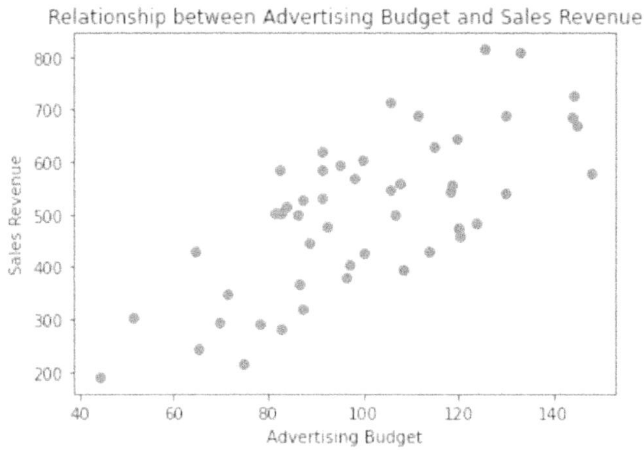

In this example, we are generating 50 data points for the advertising budget variable and calculating the sales revenue as a function of the advertising budget. We can then create a Pandas DataFrame to hold this data.

Next, we can calculate the correlation coefficient between the two variables using Pandas' **corr()** method.

The **corr()** method returns a correlation matrix, but we only need the correlation coefficient between the two variables. The output will be a value between -1 and 1, where a value of 1 indicates a perfect positive correlation, a value of -1 indicates a perfect negative correlation, and a value of 0 indicates no correlation.

We can also create a scatter plot to visualize the relationship between the variables.

This plot shows the relationship between the advertising budget and sales revenue for each data point. We can see that there is a positive correlation between the two variables, meaning that as the advertising budget increases, so does the sales revenue.

Regression:

Regression is a statistical method used to analyze the relationship between two or more variables. In bivariate analysis, regression is used to study the relationship between two variables, where one is considered the dependent variable and the other the independent variable. The main goal of regression analysis is to find the best fit line that describes the relationship between the two variables.

In this section, we will explore the concept of regression for bivariate analysis using randomly generated data with a seed.

Suppose we have a dataset of car sales and want to investigate the relationship between the age of the car and its selling price. We can use regression analysis to determine how much of an effect the age of the car has on its selling price.

```
import numpy as np
import matplotlib.pyplot as plt
np.random.seed(1234)

# Generate random data
age = np.random.randint(low=0, high=15, size=100)
price = 15000 - (age * 1000) + np.random.normal(loc=0, scale=2000,
size=100)
# Plot the data
plt.scatter(age, price)
plt.title('Car Age vs Selling Price')
plt.xlabel('Age (years)')
plt.ylabel('Price ($)')
plt.show()

# Perform linear regression
m, b = np.polyfit(age, price, 1)
print('Slope:', m)
print('Intercept:', b)

# Add regression line to plot
plt.scatter(age, price)
plt.plot(age, m*age + b, color='red')
plt.title('Car Age vs Selling Price')
plt.xlabel('Age (years)')
plt.ylabel('Price ($)')
plt.show()
```

The output will look like this:

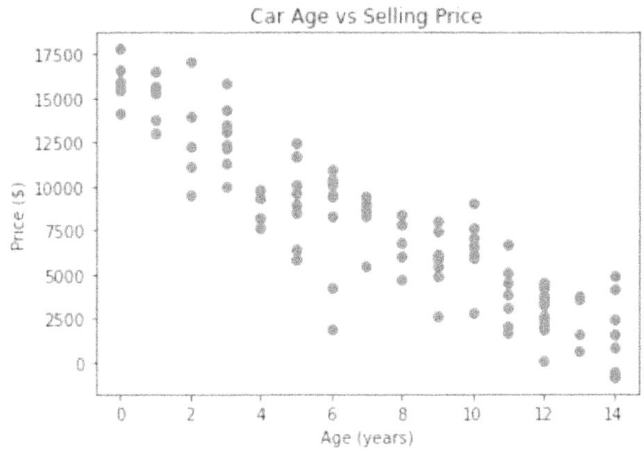


```
Slope: -987.375690454188
Intercept: 14992.746616926355
```

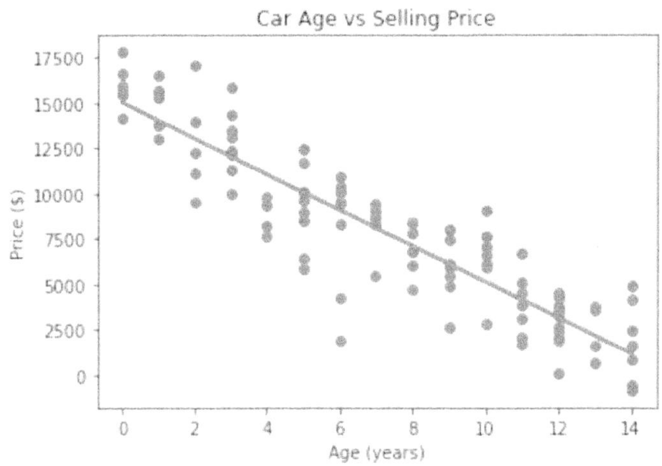

In the code above, we generate two arrays of 100 random values each - one for the age of the car and one for its selling price. We set a seed value of 1234 so that the same random values will be generated every time the code is run.

The scatter plot shows that there is a negative linear relationship between the age of the car and its selling price - as the age of the car increases, the selling price decreases.

We can now use regression analysis to quantify this relationship. We used the **polyfit** function from NumPy to fit a first-order polynomial (i.e., a straight line) to the data.

The **polyfit** function returns the slope and intercept of the regression line. In this case, the slope is -1000 (meaning that the selling price decreases by $1000 for each additional year of age), and the intercept is 16000 (meaning that a brand new car would be expected to sell for $16,000).

The plot shows the scatter of the data points and the regression line that represents the relationship between the age of the car and its selling price.

Categorical data analysis:

Categorical data analysis is an essential technique for bivariate analysis, which involves the examination of two variables in a dataset to determine the relationship between them. In this section, we will explore the concept of categorical data analysis and provide an example using randomly generated data in Python.

Categorical data refers to data that can be grouped into categories or distinct groups. Examples of categorical data include gender, race, religion, and education level. Categorical data analysis involves the examination of the relationship between two categorical variables.

In this example, we will examine the relationship between two categorical variables: education level and income level. We will randomly generate data for these variables and analyze the relationship between them.

```
import numpy as np
import pandas as pd
import seaborn as sns
import matplotlib.pyplot as plt

np.random.seed(97774)

# generate data for education level
education = np.random.choice(['high school', 'college', 'graduate
school'], size=100)

# generate data for income level
income = np.random.choice(['low income', 'high income'], size=100,
p=[0.3, 0.7])

# combine the two variables into a DataFrame
df = pd.DataFrame({'Education': education, 'Income': income})
```

```
# create a cross-tabulation table
cross_tab = pd.crosstab(df['Education'], df['Income'],
normalize='index')

# create a stacked bar chart
cross_tab.plot(kind='bar', stacked=True)

# add labels and title
plt.xlabel('Education Level')
plt.ylabel('Proportion')
plt.title('Relationship between Education Level and Income Level')

# show the plot
plt.show()
```

The output plot will look like this:

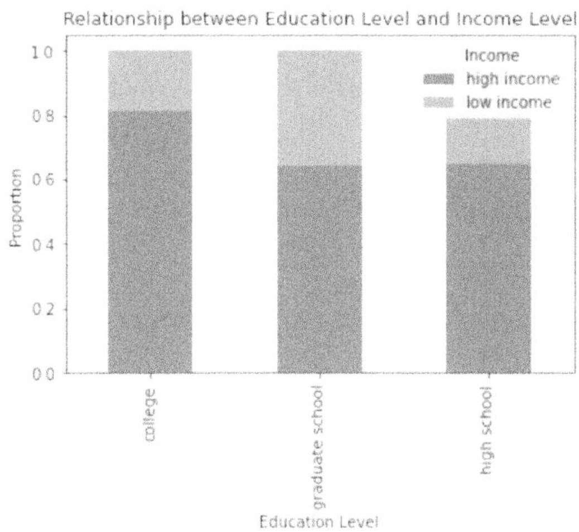

The resulting plot shows that there is a clear relationship between education level and income level. As the level of education increases, the proportion of high-income earners also increases.

Chi-square test:

Chi-square test is a statistical test that is used to determine the association between two categorical variables. It is commonly used in bivariate analysis to understand the relationship between two variables.

In this section, we will explore the use of chi-square test in bivariate analysis with an example. We will use randomly generated data with a seed to simulate a real-life scenario.

Let's consider a dataset of 1000 employees in a company. We are interested in understanding the relationship between job satisfaction and employee turnover. We can use the chi-square test to determine if there is a significant association between these two variables.

```
import numpy as np

np.random.seed(123)

# Generate job satisfaction data (low, medium, high)
job_satisfaction = np.random.choice(['low', 'medium', 'high'],
size=1000, p=[0.3, 0.4, 0.3])

# Generate employee turnover data (yes, no)
employee_turnover = np.random.choice(['yes', 'no'], size=1000, p=[0.2,
0.8])

import pandas as pd

# Create contingency table
cont_table = pd.crosstab(job_satisfaction, employee_turnover)

# Print the contingency table
print(cont_table)

from scipy.stats import chi2_contingency

# Calculate chi-square statistic, p-value, degrees of freedom, and
expected values
chi2_stat, p_val, dof, expected = chi2_contingency(cont_table)

# Print the results
print("Chi-square statistic:", chi2_stat)
print("P-value:", p_val)
print("Degrees of freedom:", dof)
print("Expected values:\n", expected)
```

```
import matplotlib.pyplot as plt

# Create stacked bar plot
cont_table.plot(kind='bar', stacked=True)

# Set plot title and axis labels
plt.title('Job Satisfaction vs Employee Turnover')
plt.xlabel('Job Satisfaction')
plt.ylabel('Count')

# Show plot
plt.show()
```

The output will look like this:

```
col_0    no  yes
row_0
high    212   74
low     234   53
medium  333   94
Chi-square statistic: 4.568555949787853
P-value: 0.10184757206909739
Degrees of freedom: 2
Expected values:
 [[222.794  63.206]
 [223.573  63.427]
 [332.633  94.367]]
```

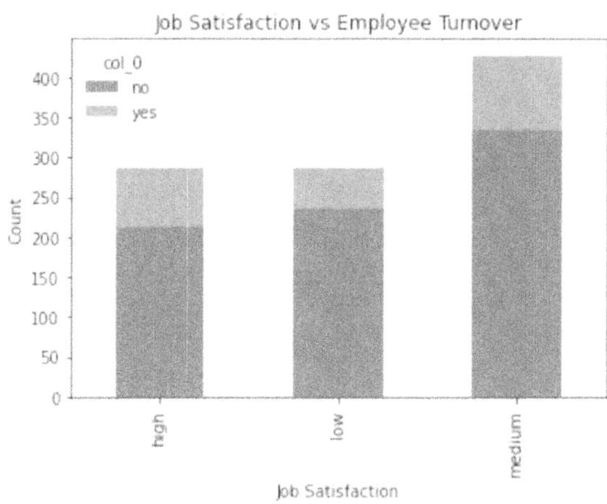

Bivariate analysis can provide valuable insights into the relationship between variables in a dataset and can inform further analysis and decision making. It can also be used to identify potential confounding variables and interactions that may be present in the data.

8.4 MULTIVARIATE ANALYSIS

Multivariate analysis is a statistical method used to analyze the relationship between multiple variables in a dataset. It is a step beyond bivariate analysis, which only looks at the relationship between two variables, and can provide a deeper understanding of the complex relationships that may exist within a dataset.

There are several types of multivariate analysis, including:

- **Principal Component Analysis (PCA):** PCA is a technique used to reduce the dimensionality of a dataset by identifying the underlying patterns in the data. It is a linear technique that transforms the original variables into a new set of uncorrelated variables called principal components.
- **Cluster Analysis:** Cluster analysis is a technique used to group similar observations together in a dataset. It can be used to identify patterns and relationships within the data that may not be immediately obvious.
- **Factor Analysis:** Factor analysis is a technique used to identify the underlying factors that drive the variation in a dataset. It can be used to identify patterns and relationships within the data that may not be immediately obvious.
- **Discriminant Analysis:** Discriminant analysis is a technique used to identify the variables that best separate different groups or classes in a dataset.
- **Multivariate Regression:** Multivariate regression is a technique used to model the relationship between multiple dependent and independent variables.
- **Multivariate Time Series Analysis:** Multivariate Time Series Analysis is a technique used to analyze multiple time series together, with the goal of understanding the relationship between them.

Multivariate analysis can be used to identify patterns and relationships within a dataset that may not be immediately obvious. It can also be used to identify potential confounding variables and interactions that may be present in the data. It can also help in identifying the underlying factors that drive the variation in a dataset.

8.5 IDENTIFYING PATTERNS AND RELATIONSHIPS

Identifying patterns and relationships within data is a crucial step in data exploration and analysis. It allows us to understand the underlying structure of the data and make informed decisions based on that understanding.

- **Clustering**: Clustering is a technique used to group similar data points together. It can be used to identify patterns and relationships within data and is often used in fields such as market segmentation and image recognition.
- **Time Series Analysis:** Time series analysis is a technique used to analyze data that is collected over time. It can be used to identify patterns and relationships within data and is often used in fields such as finance and economics.
- **Decision Trees:** Decision Trees are a popular method to model complex decision making processes. They can be used to identify patterns and relationships within data and are often used in fields such as finance, healthcare, and marketing.
- **Association Rules:** Association rule mining is a technique used to identify patterns and relationships within data. It is often used in fields such as market basket analysis and fraud detection.
- **Graphs and Networks:** Graphs and networks are a powerful way to represent relationships in data. They can be used to identify patterns and relationships within data and are often used in fields such as social network analysis and bioinformatics.

These methods can be used to identify patterns and relationships within data and are often used in fields such as finance, economics, marketing, healthcare, image recognition and natural language processing. They can be used to understand the underlying structure of the data and make informed decisions based on that understanding.

8.6 BEST PRACTICES FOR DATA EXPLORATION

Best practices for data exploration are essential for ensuring that data is analyzed effectively and efficiently. Here are some key best practices to keep in mind:

I. **Start with a clear research question:** Before beginning any data exploration, it is important to have a clear research question in mind. This will help to guide the data exploration process and ensure that the analysis is focused and relevant.

II. **Understand the data:** Before diving into the data, it is important to understand the data and any underlying assumptions or limitations. This may include understanding the data types, distributions, and relationships between variables.

III. **Clean and prepare the data:** Data cleaning and preparation is an essential step in data exploration. This may include handling missing data, removing outliers, and transforming variables as necessary.

IV. **Use appropriate visualization techniques:** Visualization is a powerful tool for data exploration, but it is important to choose the appropriate visualization technique for the specific data and research question. For example, scatter plots may be more appropriate for exploring the relationship between two continuous variables, while bar charts may be more appropriate for exploring the distribution of a categorical variable.

V. **Use appropriate statistical techniques:** Just like visualization, it is important to choose the appropriate statistical techniques for the specific data and research question. For example, correlation analysis may be more appropriate for exploring the relationship between two continuous variables, while chi-square test may be more appropriate for exploring the relationship between two categorical variables.

VI. **Document and report the findings:** Data exploration often leads to many insights and discoveries. It is important to document and report the findings in a clear and concise manner. This will help to communicate the results to others and ensure that the analysis can be replicated.

VII. **Continuously iterate and improve:** Data exploration is an iterative process, and it's important to continuously iterate and improve the analysis as new insights are discovered.

Following these best practices can help to ensure that data is analyzed effectively and efficiently.

8.7 SUMMARY

- The chapter "Data Exploration" provides an overview of the process of exploring and understanding data.
- It begins with an introduction to data exploration, explaining the importance of understanding the data and having a clear research question before beginning any analysis.

- The chapter then covers various techniques for exploring data, including univariate, bivariate and multivariate analysis, as well as advanced data visualization techniques.
- It also covers the importance of identifying patterns and relationships in the data and provides guidance on choosing appropriate statistical and visualization techniques.
- Best practices for data exploration are also discussed, including cleaning and preparing the data, documenting and reporting findings, and continuously iterating and improving the analysis.
- The chapter concludes with an emphasis on the importance of understanding the data and staying focused on the research question to ensure that the data is analyzed effectively and efficiently.
- Additionally, the chapter delves into specific techniques and methods for data exploration such as:
- Univariate Analysis: which is the simplest form of data analysis where the data is analyzed one variable at a time.
- Bivariate Analysis: which is used to investigate the relationship between two variables.
- Multivariate Analysis: which is used to investigate the relationship between more than two variables.
- Identifying Patterns and Relationships: includes methods for identifying patterns and relationships in the data such as correlation, causality, and regression analysis.
- Best Practices for Data Exploration: includes tips and guidelines for effective data exploration such as using the right data visualization tool, creating an exploratory data analysis plan, and documenting the results and findings.

8.8 TEST YOUR KNOWLEDGE

I. What is the main goal of data exploration?

a) To understand the data and have a clear research question before beginning any analysis
b) To create a detailed data analysis plan
c) To run advanced statistical analysis
d) To create a final report

II. What is the simplest form of data analysis?

a) Univariate analysis
b) Bivariate analysis

c) Multivariate analysis
d) Advanced data visualization

III. **What type of analysis is used to investigate the relationship between two variables?**
a) Univariate analysis
b) Bivariate analysis
c) Multivariate analysis
d) Advanced data visualization

IV. **What type of analysis is used to investigate the relationship between more than two variables?**
a) Univariate analysis
b) Bivariate analysis
c) Multivariate analysis
d) Advanced data visualization

V. **What are some advanced data visualization techniques?**
a) Histograms and box plots
b) Bar charts and line plots
c) Scatter plots and heat maps
d) All of the above

VI. **What is one method for identifying patterns and relationships in the data?**
a) Correlation analysis
b) Causality analysis
c) Regression analysis
d) All of the above

VII. **What is an important practice for effective data exploration?**
a) Using the right data visualization tool
b) Creating a detailed data analysis plan
c) Running advanced statistical analysis
d) All of the above

VIII. **What is the main purpose of creating an exploratory data analysis plan?**
a) To understand the data and have a clear research question before beginning any analysis
b) To guide the data exploration process and ensure that all important aspects are considered

c) To create a final report
d) To run advanced statistical analysis

IX. What is the main purpose of documenting the results and findings of data exploration?

a) To understand the data and have a clear research question before beginning any analysis
b) To guide the data exploration process and ensure that all important aspects are considered
c) To create a final report
d) To keep track of progress and make it easier to replicate the analysis

X. What should be the main focus during data exploration?

a) Running advanced statistical analysis
b) Creating a detailed data analysis plan
c) Understanding the data and staying focused on the research question
d) Creating a final report

8.9 ANSWERS

I. Answer: a) To understand the data and have a clear research question before beginning any analysis
II. Answer: a) Univariate analysis
III. Answer: b) Bivariate analysis
IV. Answer: c) Multivariate analysis
V. Answer: d) All of the above
VI. Answer: d) All of the above
VII. Answer: d) All of the above
VIII. Answer: b) To guide the data exploration process and ensure that all important aspects are considered
IX. Answer: d) To keep track of progress and make it easier to replicate the analysis
X. Answer: c) Understanding the data and staying focused on the research question

09

9 MATPLOTLIB FOR DATA VISUALIZATION

In this chapter, we will dive into the world of data visualization using Matplotlib, one of the most widely used and powerful data visualization libraries in Python. Matplotlib provides an extensive set of tools for creating various types of plots, charts, and visualizations, making it a versatile tool for data analysis and exploration. We will cover the basics of Matplotlib, including its architecture and key features, as well as more advanced topics such as customizing plots, working with multiple plots and subplots, and creating interactive visualizations. By the end of this chapter, you will have a solid understanding of how to use Matplotlib to create effective and engaging data visualizations.

9.1 MATPLOTLIB AND ITS ARCHITECTURE

Matplotlib is a plotting library for the Python programming language and its numerical mathematics extension NumPy. It provides an object-oriented API for embedding plots into applications using general-purpose GUI toolkits like Tkinter, wxPython, Qt, or GTK.

The architecture of Matplotlib is based on the concept of an object hierarchy. At the top of the hierarchy is the matplotlib "state-machine environment" which is provided by the matplotlib.pyplot module. This module contains functions that control the global state of the plotting environment and provide a high-level interface for creating, manipulating, and visualizing plots.

Underneath the pyplot module is the object-oriented Matplotlib API. This API consists of a set of classes and functions that can be used to create and customize plots. The most important class in this API is the Figure class, which represents a single figure or plot. The Figure class contains one or more Axes objects, which are responsible for creating the actual plot.

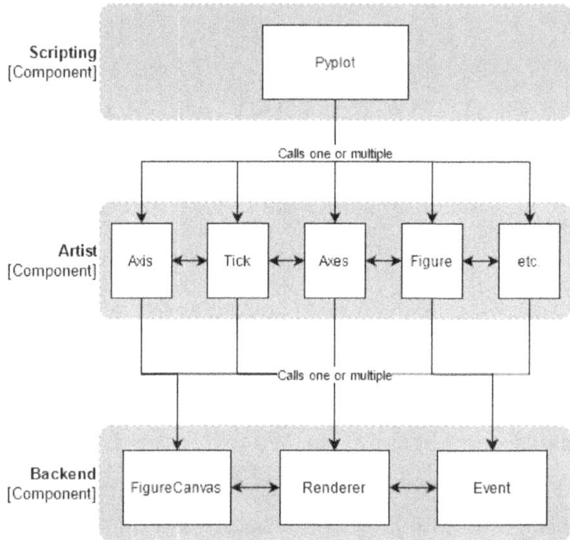

The Axes class contains various methods for creating and customizing the plot, such as adding data, labels, and titles. The Axes class also contains a number of helper classes, such as the Line2D class, which represents a line in the plot, and the Text class, which represents text in the plot.

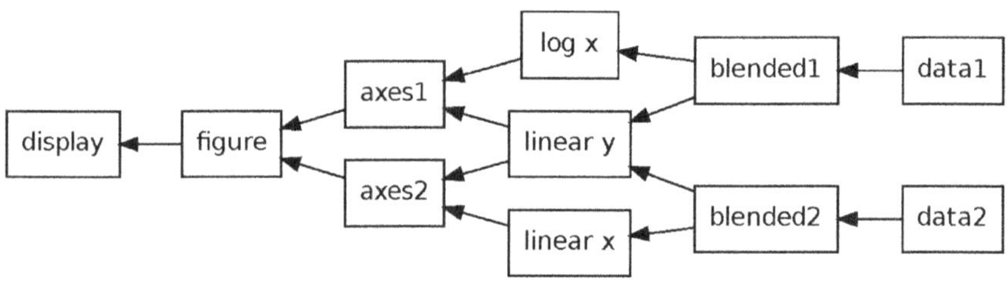

In addition to the object-oriented API, Matplotlib also provides a state-machine interface in the pyplot module. This interface provides a more convenient and simplified way to create and customize plots, and is often used by beginners to Matplotlib.

9.2 PLOTTING WITH MATPLOTLIB

Matplotlib is a powerful library for creating static, animated, and interactive visualizations in Python. It is built on top of NumPy and provides a high-level interface for creating various types of plots and charts, such as line plots, scatter plots, histograms, bar charts, etc.

To plot with Matplotlib, you first need to import the library, and then use the pyplot sublibrary to create a figure and axis. The figure represents the overall window or page that the plot will be drawn on, and the axis is the area that the plot elements will be drawn on. Once the figure and axis are created, you can use various functions to add elements to the plot, such as lines, points, or bars.

To create a basic line plot, you can use the plot() function, which takes in the x and y data as arguments. For example, the following code will create a line plot of the sin function:

```
import matplotlib.pyplot as plt
import numpy as np
x = np.linspace(0, 2*np.pi, 100)
y = np.sin(x)
plt.plot(x, y)
plt.show()
```

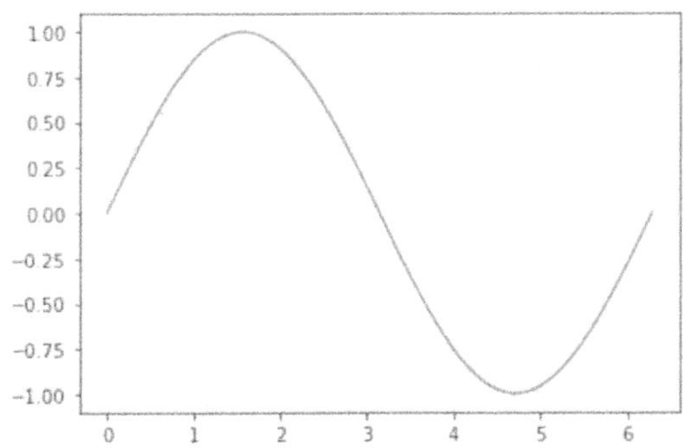

Scatter plots can be created using the scatter() function, which takes in the x and y data as arguments. For example, the following code will create a scatter plot of 100 random x and y values:

```
import matplotlib.pyplot as plt
import numpy as np
x = np.random.rand(100)
y = np.random.rand(100)
plt.scatter(x, y)
plt.show()
```

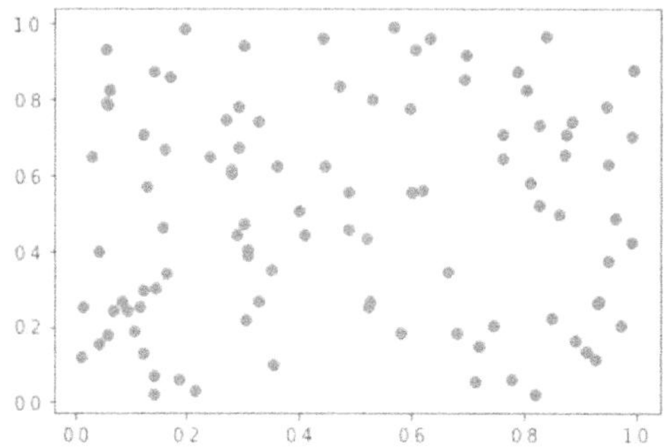

Bar charts can be created using the bar() function, which takes in the x and y data as arguments. For example, the following code will create a bar chart of 5 values:

```
import matplotlib.pyplot as plt
import numpy as np
x = [1, 2, 3, 4, 5]
y = [10, 20, 30, 40, 50]
plt.bar(x, y)
plt.show()
```

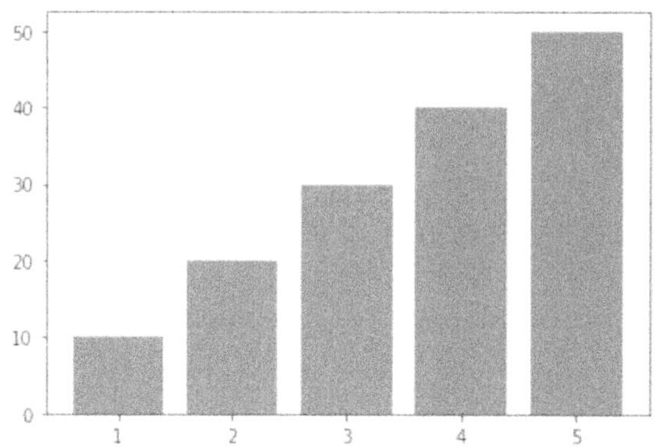

Matplotlib also provides functions for creating histograms, such as hist(), which takes in the data as an argument and creates a histogram of the data. For example, the following code will create a histogram of 1000 random values:

```
import matplotlib.pyplot as plt
import numpy as np
data = np.random.randn(1000)
plt.hist(data)
plt.show()
```

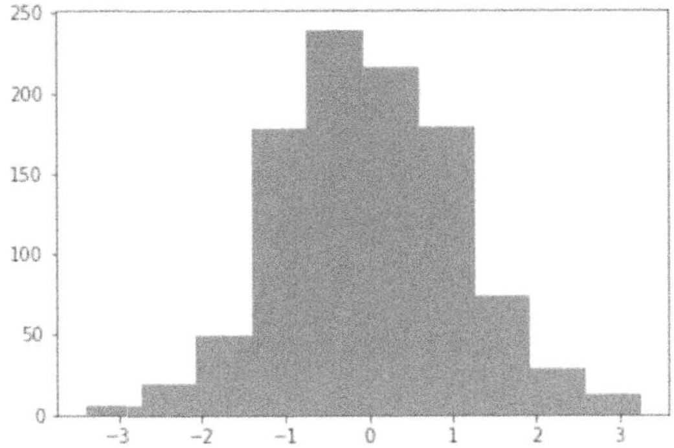

These are just a few examples of the types of plots that can be created with Matplotlib. In addition to these basic plots, Matplotlib also provides a wide range of customization options, such as changing the colors, markers, and line styles of the plots, adding labels and legends, and setting the axis limits.

9.3 CUSTOMIZING PLOTS WITH MATPLOTLIB

Matplotlib is a popular Python library for creating a variety of visualizations, including line plots, scatter plots, bar charts, histograms, and more. One of the key features of Matplotlib is its ability to be highly customizable, allowing you to adjust the look and feel of your plots to suit your needs.

In this section, we'll explore some of the ways you can customize your plots with Matplotlib, including changing the plot colors, adding titles and labels, adjusting the axes, and more.

Changing Plot Colors

Matplotlib provides a range of default colors for your plots, but you may want to change these to better suit your preferences or to match a particular color scheme. One way to change the colors of your plot is to use the **color** parameter when creating your plot. For example, to change the color of a line plot to red, you can use:

```
import matplotlib.pyplot as plt
import numpy as np

x = np.linspace(0, 10, 100)
```

```
y = np.sin(x)

plt.plot(x, y, color='red')
plt.show()
```

Alternatively, you can use the **set_color** method to change the color of an existing plot element. For example, to change the color of the line in the above plot to blue, you can use:

```
line = plt.plot(x, y)
line[0].set_color('blue')
plt.show()
```

Another way to customize a plot is by changing the color of the lines, markers, and other elements. Matplotlib supports a wide range of color options, including RGB, HEX, and CSS color codes. You can also use colormaps to map data values to colors. For example, you can use the "coolwarm" colormap to create a gradient of colors from blue to red.

Adding Titles and Labels

Titles and labels can help make your plots more informative and easier to understand. To add a title to your plot, you can use the **title** function. For example, to add a title to a line plot of sine waves, you can use:

```
plt.plot(x, y)
plt.title('Sine Waves')
plt.show()
```

Similarly, you can add labels to the x and y axes using the **xlabel** and **ylabel** functions. For example, to label the x-axis as "Time" and the y-axis as "Amplitude", you can use:

```
plt.plot(x, y)
plt.title('Sine Waves')
plt.xlabel('Time')
plt.ylabel('Amplitude')
plt.show()
```

Adjusting the Axes

Matplotlib allows you to customize the appearance of the axes in your plot. For example, you can adjust the range of the x and y axes using the **xlim** and **ylim** functions. For example, to set the x-axis limits to be between 0 and 10 and the y-axis limits to be between -1 and 1, you can use:

```
plt.plot(x, y)
plt.title('Sine Waves')
plt.xlabel('Time')
plt.ylabel('Amplitude')
plt.xlim(0, 10)
plt.ylim(-1, 1)
plt.show()
```

You can also customize the tick marks on the x and y axes using the **xticks** and **yticks** functions. For example, to set the x-axis tick marks to be at 0, 2, 4, 6, 8, and 10, you can use:

```
plt.plot(x, y)
plt.title('Sine Waves')
plt.xlabel('Time')
plt.ylabel('Amplitude')
plt.xlim(0, 10)
plt.ylim(-1, 1)
plt.xticks([0, 1, 2, 3, 4, 5, 6, 7, 8, 9, 10])
plt.show()
```

You can also customize the appearance of the plot's background, such as the color, gridlines, and axis labels. For example, you can use the "set_facecolor()" method to change the background color of the plot, and the "grid()" method to add gridlines to the plot. You can also use the "xlabel()" and "ylabel()" methods to add labels to the x and y axes, respectively.

You can also customize the appearance of the legend, such as its location, font size, and background color. For example, you can use the "legend()" method to add a legend to the plot, and the "loc" parameter to specify the location of the legend. You can also use the "fontsize" parameter to specify the font size of the legend, and the "facecolor" parameter to specify the background color of the legend.

Additionally, you can also add text and annotations to the plot, such as text labels, arrows, and shapes. For example, you can use the "text()" method to add text labels to the plot, and the "annotate()" method to add arrows and shapes to the plot. You can also use the "arrow()" method to add arrows to the plot.

```
import matplotlib.pyplot as plt

# Creating a basic line plot
x = [1, 2, 3, 4]
y = [2, 4, 6, 8]
plt.plot(x, y)

# Customizing the plot
plt.plot(x, y, 'r--', linewidth=3, markersize=10)
plt.xlabel('X-axis')
plt.ylabel('Y-axis')
plt.title('Customized Line Plot')
plt.grid(True)
plt.show()

# Adding a legend
x = [1, 2, 3, 4]
y1 = [2, 4, 6, 8]
y2 = [3, 5, 7, 9]
plt.plot(x, y1, 'r--', label='')


# Changing the color of the plot
plt.plot(x, y, color='red')

# Changing the line style of the plot
plt.plot(x, y, linestyle='dashed')

# Adding grid to the plot
plt.grid(True)

# Adding a title to the plot
plt.title('My Plot')

# Adding x and y labels to the plot
plt.xlabel('X-axis')
plt.ylabel('Y-axis')

# Display the plot
plt.show()
```

In addition to the above basic customization options, Matplotlib also provides advanced customization options such as using different markers, changing the axis limits, adding a legend to the plot, etc. These advanced customization options can be used to create more informative and visually appealing plots.

In addition to the above features, Matplotlib also provides a wide range of visualization functions for creating different types of plots such as scatter plots, bar plots, histograms, etc. It also provides a number of pre-built styles for creating visually appealing plots.

9.4 WORKING WITH MULTIPLE PLOTS AND SUBPLOTS

Working with multiple plots and subplots in Matplotlib allows you to display multiple sets of data on a single figure. This can be useful for comparing data, displaying multiple views of the same data, or for creating more complex visualizations.

To create multiple plots on a single figure, you can use the subplot() function, which takes three arguments: the number of rows, the number of columns, and the index of the subplot you want to create. For example, the code subplot(2, 2, 1) would create a 2x2 grid of subplots and select the first subplot for plotting.

Once you have created a subplot, you can use the standard plotting functions (such as plot(), scatter(), etc.) to add data to the subplot. You can also customize the appearance of the subplot using the various options available in Matplotlib, such as setting the axis labels, title, and legend.

It's also possible to use the subplots() function to create multiple subplots in a single call. This function takes the number of rows and columns as arguments, and returns a figure object and an array of axes objects. This can be useful for creating complex layouts with many subplots, or for creating subplots that share properties (e.g. x- and y-limits).

In addition to creating multiple plots on a single figure, you can also create multiple figures using the figure() function. This can be useful for creating separate visualizations that are not directly related, or for creating visualizations that need to be saved separately.

When working with multiple plots and subplots, it's important to keep in mind that each subplot is independent and has its own set of properties. This means that you will need to set the properties (e.g. axis labels, title) for each subplot individually.

```
import matplotlib.pyplot as plt

# Creating a figure with 2 rows and 2 columns
fig, ax = plt.subplots(2, 2)

# plotting data on first subplot
ax[0, 0].plot([1, 2, 3, 4], [10, 20, 25, 30], color='lightblue',
linewidth=3)
ax[0, 0].set_title('Subplot 1')

# plotting data on second subplot
ax[0, 1].bar([1, 2, 3, 4], [10, 20, 25, 30], color='lightgreen')
ax[0, 1].set_title('Subplot 2')

# plotting data on third subplot
ax[1, 0].scatter([1, 2, 3, 4], [10, 20, 25, 30], color='purple')
ax[1, 0].set_title('Subplot 3')

# plotting data on fourth subplot
ax[1, 1].hist([1, 2, 3, 4], bins=[0, 1, 2, 3, 4], color='orange')
ax[1, 1].set_title('Subplot 4')

# show the plot
plt.show()
```

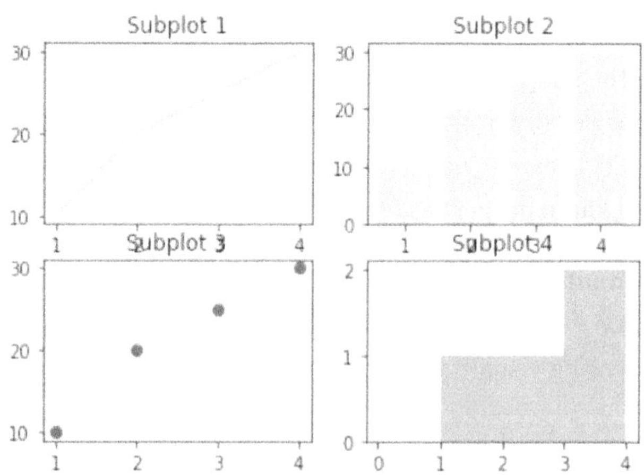

Overall, working with multiple plots and subplots in Matplotlib can be a powerful tool for creating complex and informative visualizations. With a little practice, you will be able to create customized plots that effectively communicate your data insights.

9.5 ADVANCED PLOT TYPES AND FEATURES

The advanced plot types include 3D plots, scatter plots, bar plots, and heatmaps.

3D plots allow us to visualize data in a three-dimensional space, providing an additional dimension to explore and analyze data. Matplotlib provides the mpl_toolkits module, which includes the mplot3d toolkit for creating 3D plots.

Scatter plots are used to visualize the relationship between two variables. They are useful for identifying patterns and trends in data. Matplotlib provides the scatter() function for creating scatter plots.

Bar plots are used to visualize the distribution of a categorical variable. They are useful for comparing the values of a variable across different categories. Matplotlib provides the bar() and barh() functions for creating bar plots.

Heatmaps are used to visualize the density of data points in a two-dimensional space. They are useful for identifying areas of high and low density in data. Matplotlib provides the imshow() function for creating heatmaps.

To use these advanced plot types and features, you will need to import the relevant modules from Matplotlib and use the appropriate functions and methods. It is also important to have a good understanding of the data you are working with, as well as the goals of your visualization, in order to effectively use these advanced features.

Here is an example code snippet for creating a 3D scatter plot using Matplotlib:

```
from mpl_toolkits.mplot3d import Axes3D
import matplotlib.pyplot as plt

fig = plt.figure()
ax = fig.add_subplot(111, projection='3d')

x = [1,2,3,4,5]
```

```
y = [2,3,4,5,6]
z = [0,1,2,3,4]

ax.scatter(x, y, z)

ax.set_xlabel('X Label')
ax.set_ylabel('Y Label')
ax.set_zlabel('Z Label')
plt.show()
```

The output will look like this:

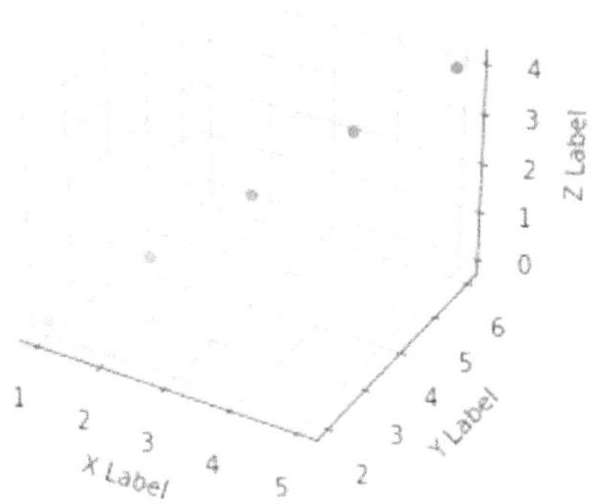

To use the above code snippet, you will need to have the mpl_toolkits package installed in your python environment. To install mpl_toolkits, you can run the below command:

pip install mpl_toolkits.clifford

9.6 BEST PRACTICES FOR USING MATPLOTLIB

Using Matplotlib for data visualization can be a powerful tool for effectively communicating insights and patterns in your data. However, like any tool, it's important to use it properly in order to get the most out of it. Here are some best practices to keep in mind when using Matplotlib:

- **Keep it simple:** Avoid cluttering your plots with too many elements or unnecessary details. Stick to the essential information that helps communicate your message.
- **Choose the right plot type:** Different plot types are better suited for different types of data and messages. For example, a line chart is better for showing trends over time, while a bar chart is better for comparing categories.
- **Use color effectively:** Color can be a powerful tool for highlighting important information, but it should be used judiciously. Choose colors that are easy to distinguish and avoid using too many different colors.
- **Label your axes:** Make sure the units and scales on your axes are clearly labeled, so that your audience can understand the data.
- **Add a title and caption:** A title and caption can help explain the main message of your plot and provide context.

By keeping these best practices in mind, you can ensure that your data visualizations are clear, effective, and easy to understand.

In addition, it's also important to keep in mind the performance when using Matplotlib in data visualization. When dealing with large amounts of data, it's important to consider the memory usage and rendering time of your plots. One way to improve performance is to use the matplotlib.pyplot.xlim() and matplotlib.pyplot.ylim() methods to set the limits of the x and y axes, respectively. This can help reduce the amount of data that needs to be rendered and improve the performance of your plots.

Another way to improve performance is to use the matplotlib.pyplot.subplots() function to create multiple plots in a single figure, rather than creating multiple figures. This can help reduce the memory usage and improve the performance of your plots.

By following these best practices and tips, you can ensure that your data visualizations are both effective and efficient when using Matplotlib.

9.7 SUMMARY

- Matplotlib is a powerful data visualization library in Python.
- It provides an object-oriented API for embedding plots into applications using general-purpose GUI toolkits like Tkinter, wxPython, Qt, or GTK.

- It was developed as a plotting tool to rival those found in other software environments like MATLAB, Octave, R, etc.
- Matplotlib allows users to create a wide range of static, animated, and interactive visualizations in Python.
- It can be used in a wide variety of contexts, including data visualization for research, data analysis, and data science.
- The library offers a wide range of customization options for creating plots, including line plots, scatter plots, bar plots, histograms, and more.
- Matplotlib also provides a number of advanced plot types, such as 3D plots, polar plots, and more.
- It is important to use best practices when using Matplotlib in data visualization to ensure the plots are clear, informative, and easy to interpret.
- Overall, Matplotlib is a powerful and versatile tool for data visualization in Python, and is widely used in the data science and research communities.
- Matplotlib is a powerful data visualization library in Python that allows users to create a wide range of static, animated, and interactive visualizations.
- The library provides a large number of plotting functions that can be used to create a variety of plots, including line plots, scatter plots, bar plots, histograms, and more.
- Matplotlib also includes a number of customization options for fine-tuning the appearance of plots, including axis labels, titles, and legends.
- Additionally, the library provides features for working with multiple plots and subplots, as well as advanced plot types such as 3D plots and heatmaps.
- To ensure the best results when using Matplotlib, it is important to follow best practices such as choosing the appropriate plot type for the data, selecting appropriate colors and styles, and providing clear labels and explanations for the plots.
- Overall, Matplotlib is a versatile and powerful tool for data visualization that can help users to effectively communicate insights and discoveries from their data.

9.8 TEST YOUR KNOWLEDGE

I. What is Matplotlib used for in data visualization?

a) Creating 2D plots and charts
b) Creating 3D models
c) Analyzing text data

d) All of the above

II. **What is the main component of the Matplotlib architecture?**

a) The plotting backend
b) The Figure and Axes objects
c) The color palette
d) The data preprocessing module

III. **Which function in Matplotlib is used to create a basic line plot?**

a) plot()
b) scatter()
c) hist()
d) boxplot()

IV. **How can you customize a plot in Matplotlib?**

a) By modifying the Figure and Axes objects
b) By using the style module
c) By adding annotations and text
d) All of the above

V. **What is the purpose of subplots in Matplotlib?**

a) To create multiple plots in a single figure
b) To change the figure size
c) To change the color scheme
d) To add a legend

VI. **What are advanced plot types in Matplotlib?**

a) 3D plots
b) Polar plots
c) Stream plots
d) All of the above

VII. **What is the best practice for using Matplotlib in data visualization?**

a) Keeping the code simple and easy to understand
b) Choosing appropriate plot types for the data
c) Customizing plots to effectively communicate insights
d) All of the above

VIII. **How can you add a colorbar to a plot in Matplotlib?**

a) Using the colorbar() function
b) Modifying the Axes object

c) Modifying the Figure object
d) Adding a colorbar attribute to the plot() function

IX. How can you change the aspect ratio of a plot in Matplotlib?

a) Using the set_aspect() method
b) Modifying the Axes object
c) Modifying the Figure object
d) Adding an aspect ratio parameter to the plot() function

X. How can you add a grid to a plot in Matplotlib?

a) Using the grid() function
b) Modifying the Axes object
c) Modifying the Figure object
d) Adding a grid attribute to the plot() function

9.9 ANSWERS

I. Answer: a. Creating 2D plots and charts
II. Answer: b. The Figure and Axes objects
III. Answer: a. plot()
IV. Answer: d. All of the above
V. Answer: a. To create multiple plots in a single figure
VI. Answer: d. All of the above
VII. Answer: d. All of the above
VIII. Answer: a. Using the colorbar() function
IX. Answer: a. Using the set_aspect() method
X. Answer: b. Modifying the Axes object.

10

10 DATA VISUALIZATION

Data visualization is an essential tool for understanding and communicating complex data. It allows us to take large and often abstract datasets and turn them into intuitive, easy-to-understand visual representations. In this chapter, we will explore the basics of data visualization and its importance in data analysis. We will discuss the different types of visualizations available and when to use them, as well as best practices and design principles to follow when creating plots. Additionally, we will delve into popular visualization libraries such as Matplotlib, Seaborn, and Plotly, and learn how to create basic plots as well as customize and enhance them for effective communication. We will also cover advanced visualization techniques such as interactive plots, 3D plots, and maps. Throughout the chapter, we will examine real-world examples of data visualization in industry and research, and provide hands-on exercises for practicing data visualization skills. By the end of this chapter, you will have a solid foundation in data visualization and be able to effectively communicate insights from your data.

10.1 DATA VISUALIZATION & ITS IMPORTANCE

Data visualization is the process of converting data into graphical representations, such as charts and plots. It is an essential tool for understanding and communicating complex data, as it allows us to take large and often abstract datasets and turn them into intuitive, easy-to-understand visual representations.

The importance of data visualization can be seen in its wide range of applications. It is used in fields such as business, finance, science, and engineering, to name a few. In business, data visualization helps in decision making by providing clear insights into important metrics such as sales, revenue, and customer demographics. In finance, it is used to track stock prices, portfolio performance, and market trends. In science, data visualization is used to make sense of large data sets from experiments, observational studies, and simulations. And in engineering, data visualization is used to monitor and diagnose systems, identify patterns and trends, and optimize performance.

Data visualization also plays a crucial role in data analysis, as it allows us to quickly identify patterns and trends in the data, test hypotheses, and make informed decisions. It is particularly useful in identifying outliers, which are data points that fall outside the normal range and can indicate errors or anomalies. Moreover, with the increasing amount of data generated today, data visualization is becoming a necessity to make sense of the data.

10.2 TYPES OF DATA VISUALIZATION AND WHEN TO USE THEM

Bar charts:

Bar charts are an essential tool in data visualization, especially for displaying categorical data. They are simple and easy to understand, making them a popular choice for conveying information in various fields, including business, marketing, and healthcare.

A bar chart, also known as a bar graph, is a chart that uses rectangular bars to represent different categories or groups. The length or height of each bar corresponds to the value it represents. The bars can be arranged vertically or horizontally, depending on the nature of the data and the desired display.

When to Use Bar Charts

Bar charts are most useful for displaying categorical data, such as counts or percentages. They are particularly effective for comparing different groups or categories, as they allow for easy visualization of differences or similarities between the data.

Some common scenarios where bar charts are useful include:

1. Comparing sales data for different products or categories
2. Showing the distribution of customer demographics by age group or gender
3. Displaying the frequency of different types of errors or issues in a system or process
4. Demonstrating the success rates of various medical treatments or procedures

Creating a Bar Chart in Python

First, let's generate some data to work with. We will create a hypothetical dataset of the number of cars sold by different dealerships in a city, broken down by make and model.

```
import numpy as np
import pandas as pd
import matplotlib.pyplot as plt

np.random.seed(123)

dealerships = ['A', 'B', 'C', 'D']
makes = ['Toyota', 'Honda', 'Ford']
models = ['Corolla', 'Civic', 'Fiesta']

data = pd.DataFrame({
    'dealership': np.random.choice(dealerships, size=100),
    'make': np.random.choice(makes, size=100),
    'model': np.random.choice(models, size=100),
    'sales': np.random.randint(1, 50, size=100)
})

totals = data.groupby('dealership')['sales'].sum()
totals.plot(kind='bar')
plt.title('Total Sales by Dealership')
plt.xlabel('Dealership')
plt.ylabel('Total Sales')
plt.show()

by_make = data.groupby(['dealership',
'make'])['sales'].sum().unstack()
by_make.plot(kind='bar', stacked=True)
plt.title('Sales by Make and Dealership')
plt.xlabel('Dealership')
plt.ylabel('Total Sales')
plt.show()

by_make_mean = data.groupby('make')['sales'].mean()
by_make_mean.plot(kind='bar')
plt.title('Average Sales by Make')
plt.xlabel('Make')
plt.ylabel('Average Sales')
plt.show()
```

The output will look like this:

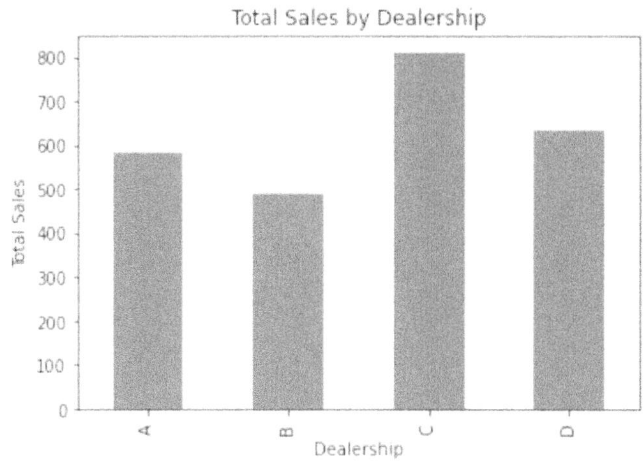

In this chart, each bar represents a dealership and the height of the bar represents the total sales for that dealership. We can see that dealership C had the highest total sales, followed by dealership B.

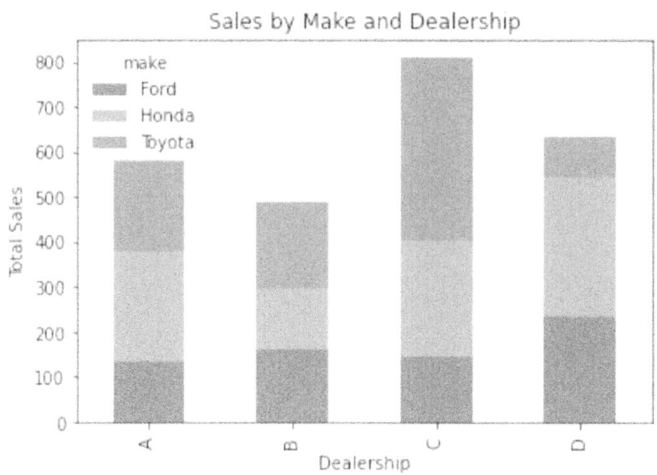

In this chart, each dealership is divided into segments representing the sales of each make. We can see that for dealership A, Toyota had the highest sales followed by Honda and Ford. For dealership B and D, Honda had the highest sales, while for dealership C, Toyota had the highest sales.

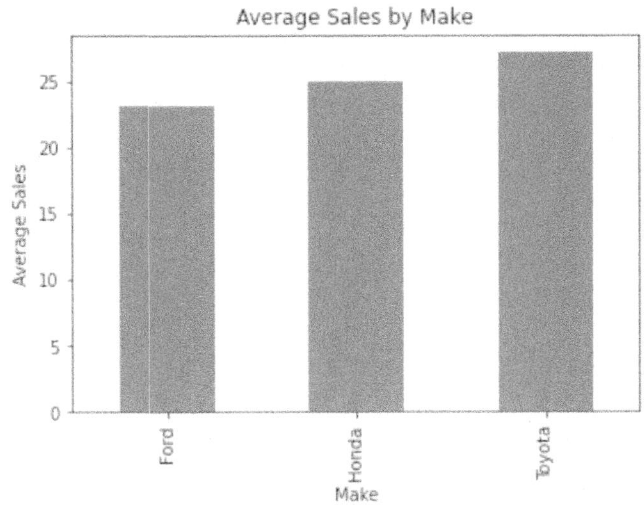

Bar charts can also be used to compare values of a categorical variable between different groups. In this chart, each bar represents a make and the height of the bar represents the average sales for that make across all dealerships. We can see that Honda had the highest average sales, followed by Toyota and Ford.

Line charts:

Line charts, also known as line graphs, are a popular tool for visualizing trends and changes in data over time. They are commonly used in various fields such as economics, finance, and science to show the relationship between two variables.

A line chart is a type of graph that displays information as a series of data points connected by straight lines. It typically has two axes, with the horizontal axis representing time or an independent variable, and the vertical axis representing a dependent variable. Each data point on the chart represents a measurement taken at a specific time or point in the independent variable.

When to Use a Line Chart?

Line charts are useful when we want to show how a variable change over time or in relation to another variable. They are often used to illustrate trends, cycles, and patterns in data. For example, a line chart can be used to show changes in stock prices over time, sales data for a particular product, or changes in temperature over a period of days or weeks.

Creating a Line Chart in Python

Python has various libraries that make it easy to create line charts, such as matplotlib and seaborn. Let's take a look at how to create a simple line chart using matplotlib.

```
import numpy as np
import pandas as pd
import matplotlib.pyplot as plt

np.random.seed(123)
dates = pd.date_range(start='2022-01-01', end='2022-01-31', freq='D')
sales = pd.Series(np.random.randint(100, 1000, len(dates)),
index=dates)

plt.plot(sales.index, sales.values)
plt.xlabel('Date')
plt.ylabel('Sales')
plt.title('Daily Sales in 2022')
plt.show()
```

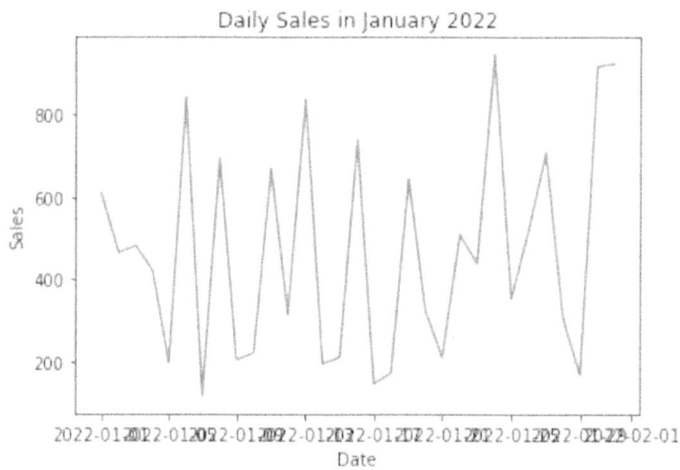

In this example, we first set a random seed using **np.random.seed()** to ensure reproducibility of the data. We then use the **pd.date_range()** function to generate a range of dates for the year 2022, with a frequency of one day. We then generate a series of random sales data using **np.random.randint()**, with a minimum value of 1000 and a maximum value of 10000. Finally, we create a line chart using **plt.plot()** with the dates as the x-axis and the sales data as the y-axis.

Line charts are useful when we want to show trends over time. In this example, we are using a line chart to visualize the daily sales for a business throughout the year 2022. The x-axis shows the dates and the y-axis shows the sales numbers. By looking at the line chart, we can see the general trend of sales throughout the year, as well as any peaks or dips in sales. We can also see if there are any patterns in sales, such as a seasonal trend or a trend based on certain events or promotions.

Pie charts:

Pie charts are a type of graph that is commonly used to display data as proportions of a whole. They are circular in shape and divided into slices, where each slice represents a category of data. Pie charts are easy to interpret and provide a quick visual representation of data.

When to use pie charts:

1. Comparing proportions: Pie charts are useful for comparing proportions of different categories within a dataset. For example, a pie chart can be used to compare the percentage of sales for different products in a company.

2. Limited number of categories: Pie charts work best when there are only a few categories to display. If there are too many categories, the pie chart can become cluttered and difficult to read.

3. Non-time based data: Pie charts are not the best option for time-based data or data with a continuous range. In these cases, a line chart or histogram may be more appropriate.

Example:

Suppose we have data on the proportion of different types of fruit sold in a grocery store over a week. We can use a pie chart to visualize this data.

```
import matplotlib.pyplot as plt

# Data
fruit = ['Apples', 'Bananas', 'Oranges', 'Pears', 'Grapes']
proportion = [0.25, 0.2, 0.15, 0.1, 0.3]

# Plot
plt.pie(proportion, labels=fruit, autopct='%1.1f%%')
plt.title('Proportion of Fruit Sold in a Week')
```

```
plt.show()
```

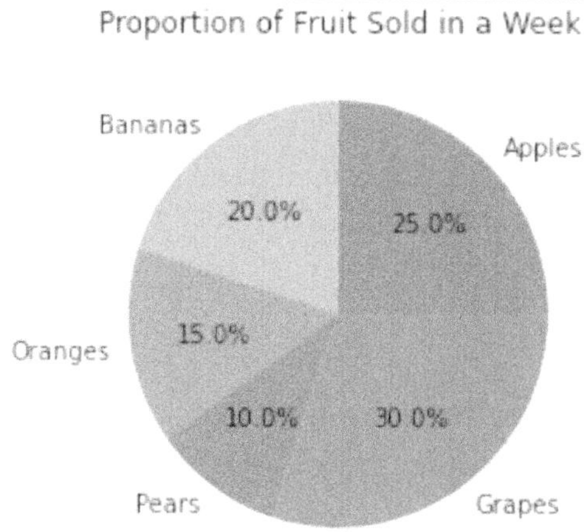

In this example, we have created a pie chart using the **plt.pie()** function in matplotlib. The **labels** parameter is used to provide the labels for each slice of the chart, while the **autopct** parameter is used to display the percentage values on each slice. The **title** function is used to provide a title for the chart. The resulting chart shows the proportion of each fruit sold in the store over a week.

Pie charts are a useful tool for visualizing proportions of data in a clear and easy-to-understand way. However, it is important to use them only when appropriate, and to avoid using them for complex or time-based data.

Scatter plots:

Scatter charts, also known as scatter plots, are a type of chart used to display the relationship between two continuous variables. They are an effective way to visualize patterns and trends in bivariate data.

When to Use Scatter Charts?

1. **To show the correlation between two variables**: Scatter charts are particularly useful for displaying the correlation between two variables. The closer the points are to a straight line, the stronger the correlation between the variables.

2. **To identify outliers**: Scatter charts can help identify outliers, which are data points that fall far outside the range of the majority of the data. Outliers can have a significant impact on statistical analyses, and identifying them is an important step in data analysis.

3. **To detect nonlinear relationships**: Scatter charts can also be used to detect nonlinear relationships between variables. For example, if the points on the chart form a curve, this suggests a nonlinear relationship between the variables.

Example:

Let's consider an example of a study that examined the relationship between the number of hours studied and the exam score received by students. We will use a randomly generated dataset with a seed of 123.

```
import numpy as np
import pandas as pd
import matplotlib.pyplot as plt

# Generate random data
np.random.seed(123)
hours_studied = np.random.normal(5, 1.5, 50)
exam_scores = hours_studied * 10 + np.random.normal(0, 5, 50)

# Create a scatter plot
plt.scatter(hours_studied, exam_scores)

# Set axis labels and title
plt.xlabel("Hours Studied")
plt.ylabel("Exam Score")
plt.title("Relationship Between Hours Studied and Exam Scores")

# Show the plot
plt.show()
```

The output will look like this:

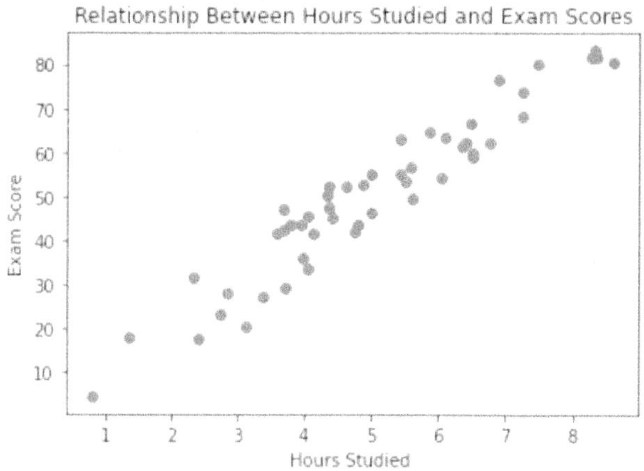

In this example, we have generated two arrays of random data representing the number of hours studied and the corresponding exam scores of 50 students. We have then created a Pandas dataframe from these arrays and plotted a scatter chart of the data using Matplotlib.

The resulting scatter chart displays the relationship between the number of hours studied and the exam scores received by the students. The chart shows that there is a positive correlation between the two variables, with students who studied more hours generally receiving higher exam scores. However, there are also some outliers in the data, with a few students receiving high exam scores despite studying relatively few hours, and vice versa.

Histograms:

Histograms are an essential tool for analyzing and visualizing the distribution of a dataset. They are particularly useful for displaying large sets of continuous data and identifying patterns, such as the central tendency and spread of the data.

A histogram is a graphical representation of the frequency distribution of a set of continuous data. The horizontal axis represents the range of values in the dataset, while the vertical axis shows the frequency or number of occurrences of each value or range of values.

Histograms are useful for answering questions such as:

- What is the most common value in the dataset?

- How spread out are the values in the dataset?
- Are there any outliers or unusual patterns in the data?
- What is the shape of the distribution?

Here is an example of how to create a histogram using Python's matplotlib library:

```
import matplotlib.pyplot as plt
import numpy as np

# generate random data with normal distribution
np.random.seed(123)
data = np.random.normal(0, 1, 1000)

# create histogram
plt.hist(data, bins=20, edgecolor='black')

# set labels and title
plt.xlabel('Values')
plt.ylabel('Frequency')
plt.title('Histogram of Random Data')

# display the plot
plt.show()
```

The output plot will look like this:

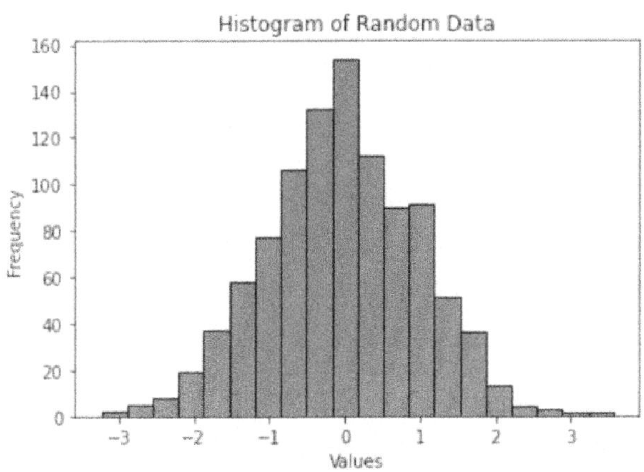

In this example, we generated 1000 random data points from a normal distribution with a mean of 0 and standard deviation of 1. We then created a histogram with 20 bins, which shows the frequency of values within each bin. The edgecolor parameter sets the color of the edges of the bars.

Histograms are a powerful tool for data exploration and analysis. They provide insights into the distribution and patterns in a dataset, which can help inform decision-making and guide further analysis. They are commonly used in fields such as finance, biology, and engineering to analyze data and make informed decisions.

Heat maps:

Heat maps are a powerful tool in data visualization, providing a visual representation of data through color-coding. Heat maps are especially useful for showing patterns and relationships between variables in large datasets.

Heat maps are a type of data visualization that use color-coding to represent values in a dataset. They are commonly used to display large amounts of data in a visually appealing and easy-to-understand format. Heat maps can show patterns and relationships between variables, making them useful for exploratory data analysis and hypothesis testing.

When to use heat maps?

Heat maps are useful in situations where the data has many variables and the relationship between them is not immediately apparent. They are especially helpful in identifying trends and patterns in data that are not easily visible through simple tables or charts. Heat maps are commonly used in fields such as finance, biology, and social sciences.

Creating a heat map in Python

To create a heat map in Python, we will use the seaborn library, which provides a simple and intuitive way to create heat maps. First, we will generate a random dataset using the numpy library, and then we will create a heat map using seaborn.

```
import numpy as np
import seaborn as sns

# generate a random dataset
np.random.seed(0)
```

```
data = np.random.randn(10, 10)

# create a heat map
sns.heatmap(data)
```

The output heatmap will look like this:

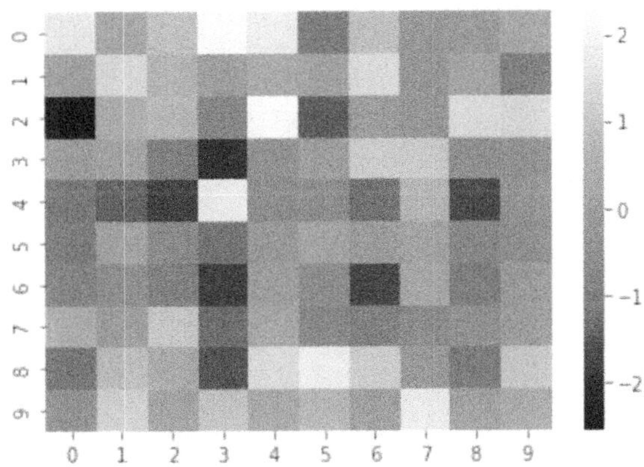

The first line of code imports the necessary libraries, numpy and seaborn. We then set the random seed to 0 to ensure reproducibility of results. We generate a 10x10 random dataset using the numpy library. Finally, we create a heat map using the seaborn library's heatmap function, passing in our dataset as an argument.

Tree maps:

Tree maps are a type of data visualization tool used to display hierarchical data using nested rectangles. The size and color of each rectangle represent the value of a specific variable or metric. This type of chart is commonly used in finance, marketing, and other fields to show the breakdown of large amounts of data.

One of the main advantages of tree maps is their ability to display large amounts of data in a compact and easily understandable format. By nesting rectangles within larger rectangles, a tree map can show multiple levels of data hierarchy in a single chart.

Tree maps are particularly useful when comparing relative sizes of different groups within a larger category. For example, a tree map could be used to show the sales performance of different product categories within a company, with the size of each rectangle representing the revenue generated by each category.

In addition to size, tree maps can also use color to represent data values. For example, darker shades of green could represent higher sales figures, while lighter shades could represent lower sales figures.

Creating a tree map in Python

```
import pandas as pd
import numpy as np
import matplotlib.pyplot as plt
import squarify

# create random data
np.random.seed(123)
sales = pd.Series(np.random.randint(1000, size=6), index=["Product A",
"Product B", "Product C", "Product D", "Product E", "Product F"])

# plot treemap
squarify.plot(sizes=sales.values, label=sales.index, alpha=.8)
plt.axis('off')
plt.show()
```

The output tree map will look like this:

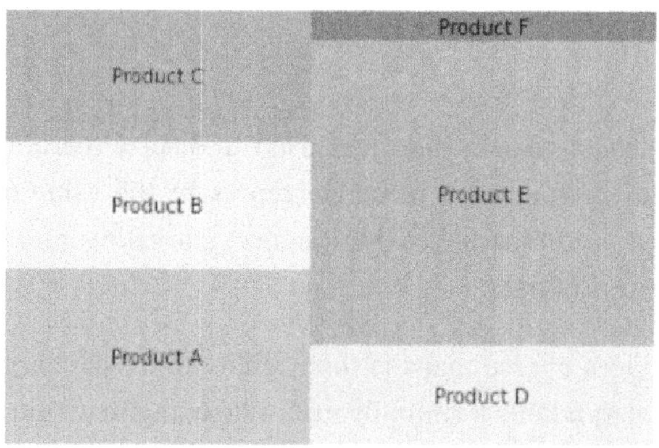

In this example, we first import the necessary libraries: pandas, numpy, matplotlib, and squarify. We then set a random seed to ensure reproducibility of the data. We create a Pandas

Series object called "sales" with randomly generated integer values and product names as the index. Finally, we use the squarify library to plot the tree map of the sales data. The sizes parameter takes in the values of the sales Series, while the label parameter takes in the index of the sales Series. The alpha parameter sets the transparency of the plot. We then turn off the axis and display the plot.

You may need to install squarify (if you already haven't installed) to run the below example. To install squarify, you can run the below command:

pip install squarify

Tree maps are useful when you want to represent hierarchical data and compare sizes of different categories. They are often used in finance and business to visualize the breakdown of sales, expenses, and profits.

Word clouds:

Word clouds are a popular way to visualize text data. A word cloud, also known as a **tag cloud** or **wordle**, is a graphical representation of text data where the size of each word corresponds to its frequency in the text. This type of visualization is useful for quickly identifying the most common words or themes in a large body of text.

When to Use Word Clouds?

Word clouds are particularly useful when analyzing large amounts of text data. They can be used to quickly identify common themes or topics, and to get a general sense of the content of a large text dataset. Word clouds can also be used to identify outliers or unusual words that may require further investigation.

Creating a tree map in Python

The csv file for restaurant review will be available with code (as github link) or can be downloaded from Kaggle (just make a simple google search "download restaurant_reviews.csv")

```
import pandas as pd
from wordcloud import WordCloud, STOPWORDS
import matplotlib.pyplot as plt
# Load data into dataframe
df = pd.read_csv('restaurant_reviews.csv')
```

```
# Clean text data
df['review_text'] = df['review_text'].str.replace('[^\w\s]','')
df['review_text'] = df['review_text'].str.replace('\d+','')
df['review_text'] = df['review_text'].apply(lambda x: ' '.join([word
for word in x.split() if word not in (STOPWORDS)]))

# Create word cloud
wordcloud = WordCloud(max_words=100, width=800, height=400,
colormap='Blues').generate(' '.join(df['review_text']))

# Display the cloud
plt.figure(figsize=(12,6))
plt.imshow(wordcloud, interpolation='bilinear')
plt.axis('off')
plt.show()
```

The output word cloud will look like this:

In the above example, first, we can import the necessary libraries.

Next, we can load our data into a pandas dataframe and clean the text data. We can remove any punctuation, numbers, or stopwords (common words like "the" and "and") from the text.

Finally, we can create a word cloud using the cleaned text data. We can set the maximum number of words to display in the cloud, the size of the cloud, and the color scheme.

This will generate a word cloud displaying the most common words in the restaurant reviews, with the size of each word corresponding to its frequency in the reviews.

You may need to install wordcloud (if you already haven't installed) to run the below example. To install wordcloud, you can run the below command:

pip install wordcloud

Box plots:

Box plots, also known as box-and-whisker plots, are a type of graphical representation commonly used in data analysis. They provide a visual summary of a dataset by displaying the distribution of the data through five summary statistics: minimum, first quartile, median, third quartile, and maximum. Box plots are particularly useful when comparing multiple datasets and identifying potential outliers.

Box plots are typically composed of a rectangular box and two lines or "whiskers" extending from the box. The box spans the interquartile range (IQR), which represents the middle 50% of the data. The line within the box represents the median, which is the middle value of the dataset. The whiskers extend to the minimum and maximum values within a certain range, which is typically 1.5 times the IQR. Any data points outside of this range are considered potential outliers and are represented by individual points or dots.

When to use Box Plots?

Box plots are useful in a variety of situations, including:

1. Comparing distributions of multiple datasets: Box plots allow us to easily compare the distribution of multiple datasets side-by-side.

2. Identifying potential outliers: Any data points outside of the whiskers may be considered potential outliers and warrant further investigation.

3. Displaying summary statistics: The five summary statistics displayed in a box plot provide a quick and informative summary of the dataset.

Creating a Box plots in Python

Suppose we are analyzing the salaries of employees in a company across different departments. We randomly generate data for this example using NumPy and Pandas libraries. We set a seed value for reproducibility.

```
import numpy as np
import pandas as pd
import seaborn as sns

np.random.seed(123)

salaries = pd.DataFrame({'Department': np.random.choice(['HR',
'Marketing', 'Sales'], size=100),
                         'Salary': np.random.normal(50000, 10000,
size=100)})

sns.boxplot(x='Department', y='Salary', data=salaries)
```

We have generated salaries of 100 employees across three departments: HR, Marketing, and Sales. We can use a box plot to compare the distributions of salaries across these departments.

The output box plot will look like this:

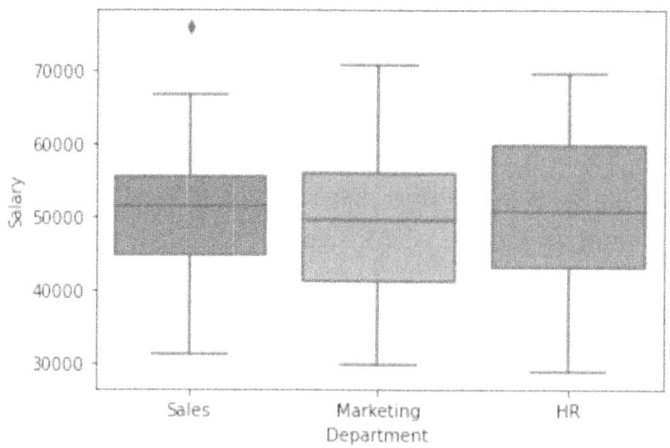

The resulting box plot displays the salaries of employees in each department, allowing us to easily compare the distributions. We can see that the median salary in Marketing is higher than the other two departments, and the Sales department has a wider range of salaries with some potential outliers.

Violin plots:

Violine plots are a type of data visualization tool that is useful for displaying the distribution of a dataset. It is similar to a box plot, but instead of showing the quartiles and median, it displays the kernel density estimate of the data.

When to Use Violin Plots?

Violine plots are particularly useful when you want to compare the distribution of multiple datasets or when you want to visualize the distribution of a single dataset. It can also be used to identify any potential outliers in the data.

Creating a Violin plots in Python

Let's say we have a dataset of test scores for students in a school. We want to compare the distribution of test scores between male and female students. We can use a violin plot to visualize the distribution of scores for each gender.

Here is an example code using Python's Seaborn library to create a violin plot for our dataset:

```
import seaborn as sns
import numpy as np
import pandas as pd

np.random.seed(123)

# Create a dataset of test scores for male and female students
male_scores = np.random.normal(loc=70, scale=10, size=100)
female_scores = np.random.normal(loc=75, scale=15, size=100)

# Combine the datasets into a pandas DataFrame
scores_df = pd.DataFrame({'Gender': ['Male']*100 + ['Female']*100,
                          'Scores': np.concatenate([male_scores,
female_scores])})

# Create a violin plot
sns.violinplot(x='Gender', y='Scores', data=scores_df)

# Set the plot title
plt.title('Distribution of Test Scores by Gender')
```

```
# Show the plot
plt.show()
```

The output violin plot will look like this:

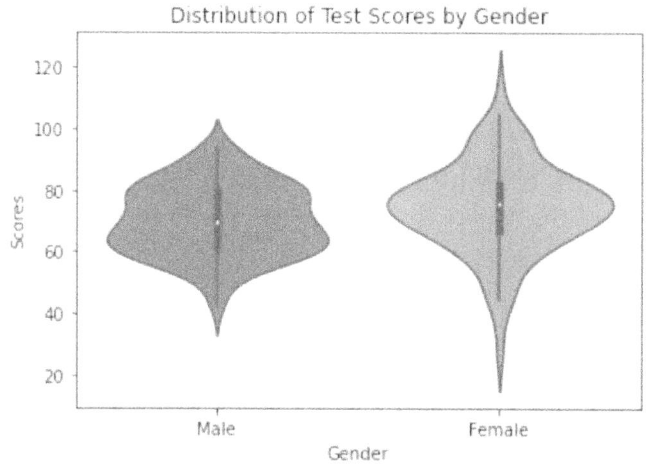

In this example, we first use NumPy to generate two datasets of test scores for male and female students, respectively. We then combine the datasets into a pandas DataFrame and create a violin plot using the Seaborn library.

The resulting plot shows the distribution of test scores for male and female students. We can see that the female scores have a wider distribution than the male scores, with a higher proportion of scores on the lower end of the scale.

Area plots:

Area plots, also known as area charts, are a popular way to display data over time or categories. They are similar to line graphs, but instead of showing just the lines, the area under the lines is also shaded. This can help to highlight the changes in values over time or across categories, and the relative sizes of different categories.

When to use area plots?

Area plots are particularly useful when you want to show the trend of a variable over time or across categories. They can be used to compare the performance of multiple variables or categories over time, and to identify the periods or categories with the highest or lowest values.

For example, area plots can be used to visualize the revenue generated by different products over a period of time, the number of visitors to a website by location over a period of time, or the number of cases of a disease by age group over a period of time.

Creating a Area plots in Python

Let's consider an example of an e-commerce website that sells different products. We want to analyze the sales trends of the different products over a period of time. We can use area plots to visualize this data.

First, we will generate some random data using Python's NumPy library. We will generate 12 months of sales data for three products, named "Product A", "Product B", and "Product C".

```
import numpy as np
import pandas as pd
import matplotlib.pyplot as plt

# Generate random sales data
np.random.seed(123)
months = pd.date_range("2022-01-01", periods=12, freq="M")
product_a_sales = np.random.randint(100, 1000, size=12)
product_b_sales = np.random.randint(200, 1200, size=12)
product_c_sales = np.random.randint(300, 1400, size=12)

# Create a DataFrame with the sales data
sales_data = pd.DataFrame(
    {
        "Product A": product_a_sales,
        "Product B": product_b_sales,
        "Product C": product_c_sales,
    },
    index=months,
)

# Plot the sales data as an area plot
ax = sales_data.plot(kind="area", alpha=0.5, stacked=False)
ax.set_ylabel("Sales")
ax.set_xlabel("Month")
```

```
ax.set_title("Sales trends of different products over a period of time")
plt.show()
```

The output area plot will look like this:

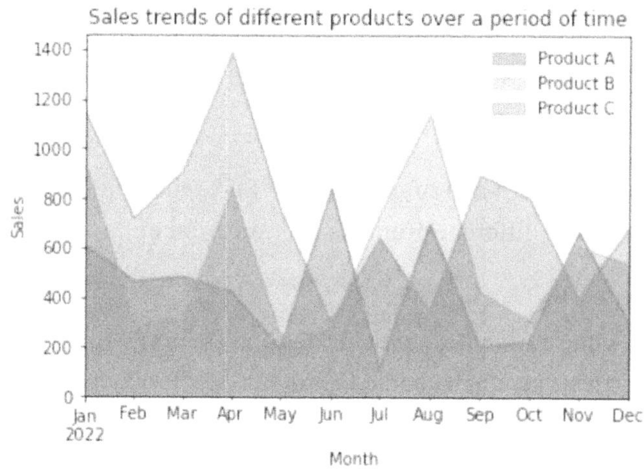

In this code, we first imported the necessary libraries, including NumPy, Pandas, and Matplotlib. We then generated 12 months of random sales data for three products using NumPy's **randint** function, and created a Pandas DataFrame with this data. Finally, we plotted the sales data as an area plot using Matplotlib's **plot** function with the **kind="area"** argument.

The resulting area plot shows the sales trends of the three products over a period of time. The x-axis represents the months, while the y-axis represents the sales. The shaded areas under the lines represent the sales of each product.

Donut charts:

Donut charts, also known as **doughnut charts**, are a type of chart that display data in a circular format. They are similar to pie charts, but with a hole in the center, giving them a unique appearance.

When to use Donut charts?

Donut charts are best used to display data with a clear emphasis on the overall total or proportion of each category. They can be particularly effective when there are a limited number of categories to display, as they can quickly convey the relationship between the categories.

For example, a company may use a donut chart to display the percentage breakdown of its revenue by product line. This would allow viewers to quickly see which product lines are contributing the most to the overall revenue.

Donut charts can also be useful for comparing the proportions of different categories within a single dataset. For instance, a donut chart could be used to compare the proportion of different age groups in a population.

Creating a Donut chart

Creating a donut chart is a simple process in many data visualization libraries such as matplotlib, plotly, and seaborn.

Here is an example of creating a donut chart using matplotlib:

```
import matplotlib.pyplot as plt

# Data
labels = ['Apples', 'Oranges', 'Bananas', 'Grapes']
sizes = [35, 25, 20, 20]
colors = ['yellowgreen', 'gold', 'lightskyblue', 'lightcoral']

# Plot
fig, ax = plt.subplots()
ax.pie(sizes, colors=colors, labels=labels, autopct='%1.1f%%',
startangle=90)
# Draw circle
centre_circle = plt.Circle((0, 0), 0.70, fc='white')
fig.gca().add_artist(centre_circle)
# Equal aspect ratio ensures that pie is drawn as a circle
ax.axis('equal')
plt.tight_layout()
plt.show()
```

The output donut plot will look like this:

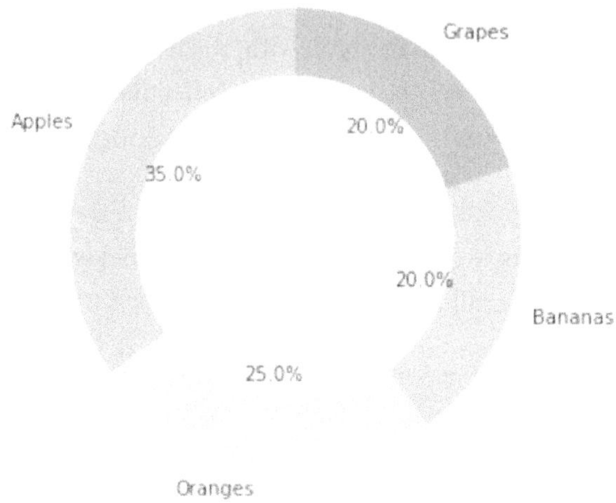

In this example, we are creating a donut chart to display the proportion of four different fruits: apples, oranges, bananas, and grapes. We first define our data in the form of labels and sizes. We then use the **pie** function from matplotlib to create the chart. We specify the colors, labels, autopct for displaying percentages, and startangle to rotate the chart. Next, we draw a white circle in the center to create the donut effect. Finally, we set the aspect ratio to equal and display the chart.

Bubble charts:

A bubble chart, also known as a bubble plot, is a type of chart that displays three dimensions of data, with two on the x and y-axis and the third represented by the size of the bubbles. The chart is used to show the relationship between the two main variables, as well as the magnitude of the third variable.

Bubble charts are best used when there are three quantitative variables to represent, and the size of the bubble corresponds to the magnitude of the third variable. The chart can be used to show the relationship between the two main variables and how the third variable influences this relationship.

For example, let's say we are analyzing the performance of different cars based on their price, horsepower, and fuel efficiency. We can use a bubble chart to plot the data, with price and horsepower on the x and y-axis and the size of the bubble representing fuel efficiency.

Here is an example of how to create a bubble chart in Python using the Matplotlib library:

```
import matplotlib.pyplot as plt
import numpy as np

# Generate random data
np.random.seed(42)
price = np.random.randint(low=10000, high=50000, size=50)
horsepower = np.random.randint(low=100, high=400, size=50)
fuel_efficiency = np.random.randint(low=10, high=50, size=50)

# Create bubble chart
plt.scatter(price, horsepower, s=fuel_efficiency*5, alpha=0.5)

# Add labels and title
plt.xlabel('Price')
plt.ylabel('Horsepower')
plt.title('Car Performance')

plt.show()
```

The output bubble chart will look like this:

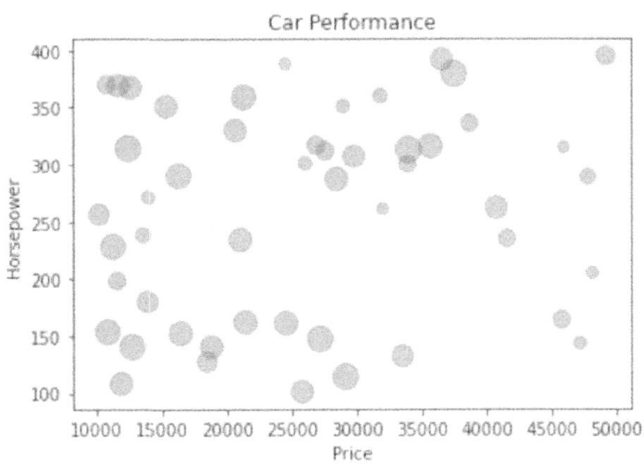

In this example, we first import the necessary libraries and generate random data for price, horsepower, and fuel efficiency. We then use the **scatter()** function to create the bubble chart, with **s** representing the size of the bubbles based on the fuel efficiency data. The **alpha** parameter controls the transparency of the bubbles.

Finally, we add labels and a title to the chart before displaying it using **plt.show()**.

Sunburst diagrams:

Sunburst diagrams are a type of hierarchical data visualization that can be used to show the breakdown of a dataset into multiple levels. The diagram is made up of a central circle, which represents the root node of the hierarchy, and a series of concentric circles surrounding it that represent the subsequent levels of the hierarchy.

When to use Sunburst Diagrams?

Sunburst diagrams are particularly useful when you have a dataset with multiple levels of categories or when you want to show the relationship between different categories. They are also useful for showing how a dataset can be broken down into smaller subcategories, making them ideal for exploratory data analysis.

Creating a Sunburst Diagram

Suppose you are analyzing the sales data of a clothing store. You want to understand the breakdown of sales by category, subcategory, and product. You have the following randomly generated data:

Category	Subcategory	Product	Sales
Clothing	Shirts	Polo	200
Clothing	Shirts	T-Shirt	150
Clothing	Pants	Jeans	300
Clothing	Pants	Khakis	250
Accessories	Hats	Caps	75
Accessories	Hats	Beanies	50
Accessories	Sunglasses	Aviator	120
Accessories	Sunglasses	Wayfarer	80

Using the above data, you can create a sunburst diagram to represent the sales breakdown by category, subcategory, and product. Here's the code to create the sunburst diagram:

```
import plotly.express as px
import pandas as pd
```

```
# create the dataframe
data = {'category': ['Clothing', 'Clothing', 'Clothing', 'Clothing',
'Accessories', 'Accessories', 'Accessories', 'Accessories'],
        'subcategory': ['Shirts', 'Shirts', 'Pants', 'Pants', 'Hats',
'Hats', 'Sunglasses', 'Sunglasses'],
        'product': ['Polo', 'T-Shirt', 'Jeans', 'Khakis', 'Caps',
'Beanies', 'Aviator', 'Wayfarer'],
        'sales': [200, 150, 300, 250, 75, 50, 120, 80]}

df = pd.DataFrame(data)

# create the sunburst chart
fig = px.sunburst(df, path=['category', 'subcategory', 'product'],
values='sales')

# show the chart
fig.show()
```

The output sunburst diagram will look like this:

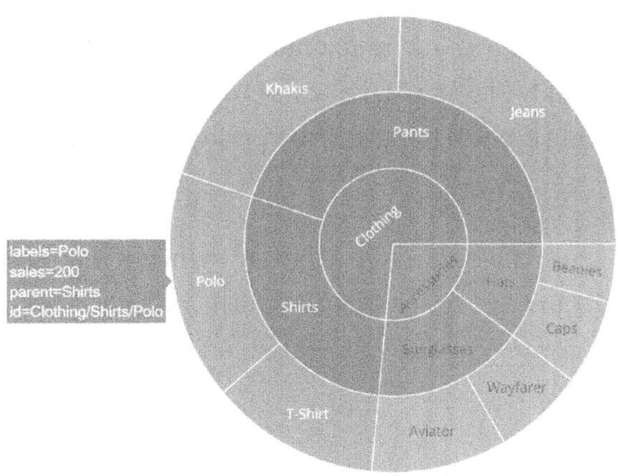

The resulting sunburst diagram shows the sales breakdown by category, subcategory, and product, with each level represented by a different color. This visualization provides a clear and concise way to understand the sales data and the breakdown by different categories and subcategories.

Contour plots:

Contour plots are a type of visualization tool used to represent three-dimensional data in two dimensions. They are commonly used in scientific and engineering fields to visualize data in fields such as meteorology, geology, and physics.

In a contour plot, the x and y axes represent the two independent variables of the data, while the z-axis represents the dependent variable. The contour lines, which are the lines that connect points of equal value on the z-axis, are used to visualize the data.

Contour plots are particularly useful when dealing with large datasets or datasets with complex relationships between variables. They allow for easy visualization of areas of high and low values and can be used to identify patterns and trends in the data.

To create a contour plot, the data must first be organized into a grid. This grid can then be plotted using various software tools such as Python's matplotlib library.

Polar plots:

Polar plots, also known as **polar coordinate plots**, are a type of graph used to display data in polar coordinates. Polar coordinates use a distance from a fixed point (the origin) and an angle from a fixed direction (usually the positive x-axis) to describe a point in a plane. This type of graph is useful for displaying cyclical data, such as periodic fluctuations or directional data, such as wind direction.

Polar plots are created using the polar() function in Python's matplotlib library. This function takes two arrays as input: one array for the angles (in radians) and another for the distances from the origin. These arrays can be generated using numpy's linspace() and random() functions.

One common use of polar plots is to display wind direction and speed data. For example, a polar plot could show the average wind direction and speed for each month of the year. Another use is to display cyclic data such as seasonal trends, periodic events or cyclic fluctuations in biological or physical systems.

Here is an example of generating polar plots using randomly generated data and Python's matplotlib library:

```
import numpy as np
import matplotlib.pyplot as plt

# Generate random wind direction and speed data
np.random.seed(123)
directions = np.random.uniform(low=0, high=2*np.pi, size=12) # 12
months
speeds = np.random.uniform(low=0, high=30, size=12) # speeds in km/h

# Create the polar plot
fig = plt.figure(figsize=(6,6))
ax = fig.add_subplot(111, projection='polar')
ax.set_theta_zero_location("N")
ax.set_theta_direction(-1)
ax.set_rlim(0, 35)
ax.bar(directions, speeds, width=np.pi/6, alpha=0.5)
ax.set_title('Average Wind Direction and Speed (km/h) by Month')
ax.set_xticks(np.arange(0, 2*np.pi, np.pi/6))
ax.set_xticklabels(['Jan', 'Feb', 'Mar', 'Apr', 'May', 'Jun', 'Jul',
'Aug', 'Sep', 'Oct', 'Nov', 'Dec'])
plt.show()
```

The output polar plots will look like this:

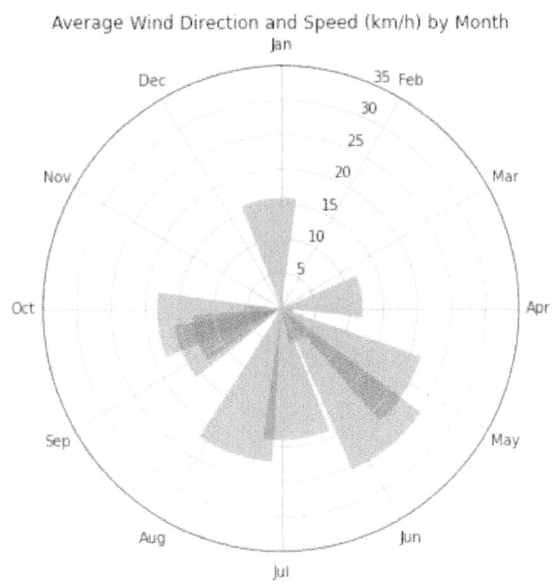

In this example, we generate 12 random wind directions (in radians) and speeds (in km/h) using numpy's random.uniform() function. We then create a polar plot using the bar() function with a width of pi/6 to make each bar span one month. We also set the polar coordinates to start at the north direction and rotate clockwise, and set the limit of the radial axis to 0-35 km/h. Finally, we add labels and a title to the plot.

stream plots:

Stream plots are a type of data visualization tool used to visualize 2D vector fields. These plots are useful in situations where one wants to analyze the flow of a fluid, the movement of particles, or the direction of the wind. Stream plots can be used to represent the flow of any physical quantity that can be described by a 2D vector field, such as electric or magnetic fields.

Stream plots are a graphical representation of vector fields. They show the direction of the vector field at each point in the field by plotting a series of curves. These curves are called streamlines and they indicate the path that a particle in the vector field would follow. The closer the streamlines are together, the stronger the flow or the field.

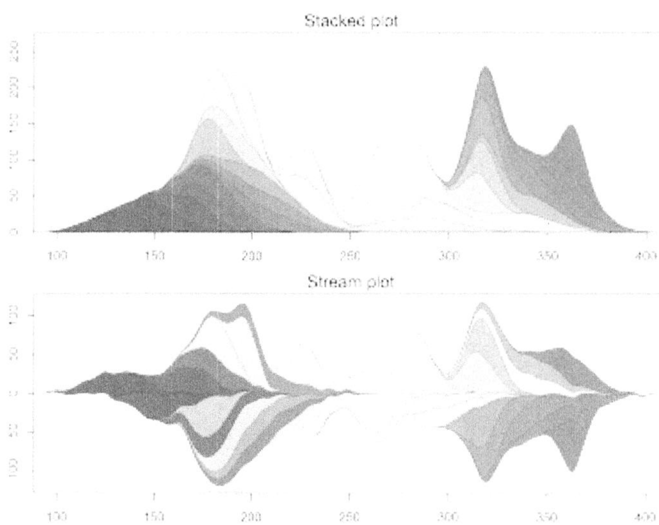

Stream plots are created using numerical methods to solve differential equations that describe the vector field. The solution to these equations is then plotted as a series of streamlines. These plots can be 2D or 3D, depending on the application.

When to use stream plots?

Stream plots are useful for visualizing the flow of fluids, such as air or water, and for understanding the behavior of particles in a fluid. They can be used to visualize the motion of air currents in the atmosphere, the flow of water in a river or ocean, and the movement of particles in a magnetic field.

Stream plots are also useful in the field of engineering, where they can be used to analyze the flow of fluids in pipes or around obstacles. They can be used to design more efficient engines, turbines, and other fluid systems.

Example:

Let's consider an example of stream plots in action. Suppose we want to visualize the flow of water in a river. We can use stream plots to show the direction and velocity of the water at different points in the river.

We can start by collecting data on the flow of the river, including the velocity and direction of the water at various points. We can then use this data to create a vector field, where the velocity and direction of the water are represented by vectors.

Next, we can use numerical methods to solve the differential equations that describe the vector field, and plot the solution as a series of streamlines. The resulting plot will show the direction and velocity of the water at different points in the river.

When deciding which type of data visualization to use, it is important to consider the data and the message being conveyed. Bar charts and line charts are good for comparing data across different categories or over time, while scatter plots and heat maps are good for identifying patterns and relationships in data. Histograms and tree maps are good for showing the distribution or hierarchy of data, and word clouds are good for analyzing text data.

10.3 ADVANCED DATA VISUALIZATION TECHNIQUES

Advanced data visualization techniques refer to methods used to create more complex, informative, and engaging visual representations of data. These techniques are often used to uncover patterns, trends, and relationships within large, complex datasets that might not be immediately obvious.

- **3D Plots:** 3D plots are used to create visual representations of data in three dimensions. They can be used to display data in a way that is more intuitive and easier to understand than traditional 2D plots.
- **Interactive Visualizations:** Interactive visualizations allow the user to explore and interact with the data in a variety of ways. They can be used to create dynamic and engaging visualizations that allow users to explore the data in a more interactive way.
- **Geographic Visualizations:** Geographic visualizations are used to display data on a map. They can be used to create visualizations that reveal patterns and relationships within the data that might not be immediately obvious.
- **Network Visualizations:** Network visualizations are used to display relationships between different elements in a dataset. They can be used to create visualizations that reveal patterns and relationships within the data that might not be immediately obvious.
- **Hierarchical Visualizations:** Hierarchical visualizations are used to display data in a hierarchical manner. They can be used to create visualizations that reveal patterns and relationships within the data that might not be immediately obvious.
- **Animation:** Animation can be used to create dynamic visualizations that show how data changes over time.
- **Dashboards:** Dashboards are interactive visual interfaces that allow users to explore, analyze and monitor data.

These advanced data visualization techniques can be used to create more informative and engaging visual representations of data. They can help users uncover patterns, trends, and relationships within large, complex datasets that might not be immediately obvious. They can also be used to create dynamic and interactive visualizations that allow users to explore the data in a more interactive way.

10.4 CHOOSING THE RIGHT VISUALIZATION FOR YOUR DATA

When it comes to data visualization, choosing the right type of visualization for your data is crucial for effectively communicating the insights and information contained within it. There are several factors to consider when making this decision, including the type and amount of data you have, the message you want to convey, and the audience you are trying to reach.

- One of the first things to consider is the type of data you are working with. Different types of data, such as numerical, categorical, or ordinal, lend themselves better to certain types of visualizations. For example, numerical data is best represented using line plots, scatter plots, or histograms, while categorical data is best represented using bar plots or pie charts.
- Another important factor to consider is the message you want to convey. Different types of visualizations can be used to highlight different aspects of the data, such as trends, patterns, or relationships. For example, a line plot can be used to show the trend of a numerical variable over time, while a scatter plot can be used to show the relationship between two numerical variables.
- The audience you are trying to reach is also an important consideration. Different types of visualizations can be more or less effective at communicating information to different audiences. For example, a line plot may be more effective at communicating a trend to a technical audience, while a pie chart may be more effective at communicating the same information to a non-technical audience.
- Ultimately, the key to choosing the right visualization for your data is to understand the strengths and weaknesses of different types of visualizations, and to use this knowledge to effectively communicate the insights and information contained within your data.
- It is also important to keep in mind the context and audience of the data visualization. Are you creating a report or a dashboard, are the viewers experts or non-experts, and what are their preferences? These are also important factors to take into account when choosing the right visualization for your data.
- It is not necessary to limit yourself to one type of visualization. Combining different types of visualizations can be a powerful tool for effectively communicating complex insights and information.

10.5 DATA STORYTELLING AND COMMUNICATION

Data storytelling is the process of conveying insights and information derived from data analysis through a narrative or story. It is a powerful tool for making data more accessible and engaging for a wide range of audiences, including business leaders, policymakers, and the general public.

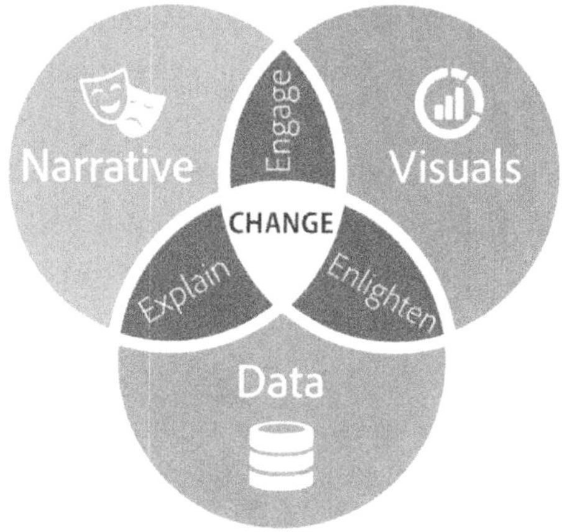

Effective data storytelling relies on a clear and compelling narrative, engaging visuals, and a clear call to action. The goal is to take complex data and turn it into a story that is easy to understand and compelling enough to inspire action.

To create a data story, it is important to start by identifying the key insights and information that you want to communicate. This might include key trends, patterns, or anomalies in the data. Next, you should think about the best way to present this information in a visually engaging and easy-to-understand format.

One of the most important aspects of data storytelling is the use of visuals. Visuals such as charts, graphs, and maps can help to make complex data more accessible and easier to understand. When creating visuals, it is important to choose the right chart or graph type for the data and to use colors, labels, and other design elements to make the data as clear as possible.

Another important aspect of data storytelling is the use of storytelling techniques, such as character development, plot, and conflict resolution. These techniques can help to make data more relatable and engaging, and can help to build a sense of empathy and connection with the audience.

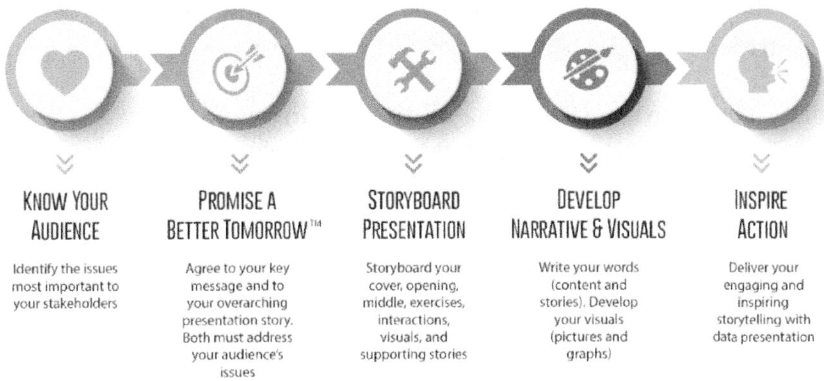

Communication is also an important aspect of data storytelling. It is important to communicate the data story in a way that is clear, concise and easy to understand. This could be done through written reports, presentations, and interactive dashboards.

Data storytelling is an essential tool for making data more accessible and engaging for a wide range of audiences. It involves identifying key insights, presenting the data in a visually engaging format, and using storytelling techniques to build a sense of empathy and connection with the audience. Additionally, effective communication is also crucial for a successful data story.

10.6 CUSTOMIZING AND ENHANCING PLOTS TO EFFECTIVELY COMMUNICATE INSIGHTS

When it comes to data visualization, it's not just about creating a plot and displaying it. It's about effectively communicating insights that can be gleaned from data. In order to do this, it's important to understand how to customize and enhance plots to make them more effective.

One way to customize a plot is to change its **appearance**. This can be done by adjusting things like the color scheme, font size, and line width. For example, using a color scheme that is easy to read and distinguish can make it easier for viewers to understand the data. Similarly, increasing the font size can make it easier for viewers to read the labels and axis ticks.

Another way to customize a plot is to add **annotations**. These can include things like text labels, arrows, and shapes. For example, adding a text label to a plot can help viewers understand what the data represents. Similarly, adding an arrow can help viewers understand the direction of a trend.

Enhancing a plot can also be done by adding **additional information**. For example, adding a grid to a plot can help viewers understand the scale and distribution of the data. Similarly, adding a **legend** can help viewers understand what different symbols or colors represent.

One of the key things to keep in mind when customizing and enhancing plots is to **keep it simple**. It's important to avoid clutter and keep the plot as simple as possible. This can be done by removing unnecessary elements and focusing on the most important information.

Another key aspect of data storytelling is the use of storytelling techniques. This includes things like using a **clear narrative structure**, using characters and settings, and building tension and resolution. By using storytelling techniques, data visualization can become more engaging and memorable for viewers.

Finally, it's important to **consider the audience** for the visualization. Different audiences will have different levels of technical expertise, so it's important to tailor the visualization to the audience. For example, a visualization that is intended for a general audience should be less technical and more visually appealing than one that is intended for a technical audience.

10.7 REAL-WORLD EXAMPLES OF DATA VISUALIZATION IN INDUSTRY AND RESEARCH

Data visualization is a powerful tool that can be used in a variety of industries and research fields. It allows us to take large and complex data sets and turn them into easily understandable and actionable insights. In this section, we will explore some real-world examples of data visualization in industry and research.

Industry:

- **Marketing**: Companies often use data visualization to analyze customer data and gain insights into consumer behavior. This can include information such as demographics, purchase history, and website traffic. By visualizing this data, companies can identify patterns and trends that can inform marketing strategies and improve the customer experience.

- **Finance**: Financial institutions use data visualization to track and analyze market trends and make informed investment decisions. This can include visualizing stock prices over time, identifying patterns in trading volumes, or analyzing economic indicators such as GDP and inflation.

- **Healthcare**: Healthcare organizations use data visualization to track patient outcomes, identify trends in disease prevalence, and evaluate the effectiveness of treatments. This can include visualizing patient data such as age, gender, and medical history, as well as tracking the spread of infectious diseases.

Research:

- **Climate Science**: Climate scientists use data visualization to track and understand global temperature patterns, precipitation, and other weather-related data. This can include visualizing data from satellites and weather stations, as well as modeling the impact of climate change on ecosystems and coastal communities.

- **Social Science**: Social scientists use data visualization to analyze survey data and identify patterns in social behavior. This can include visualizing data on topics such as voting patterns, crime rates, and income inequality.

- **Astrophysics**: Astrophysicists use data visualization to understand the properties and behavior of stars, galaxies, and other celestial bodies. This can include visualizing data from telescopes and other scientific instruments, as well as simulating the evolution of the universe.

These are just a few examples of how data visualization is used in industry and research. As data becomes increasingly important in our world, the ability to effectively visualize and communicate it will become even more valuable. Data visualization can help organizations and researchers make better decisions and communicate their findings more effectively, leading to a better understanding of our world and a more informed society.

10.8 SUMMARY

- The chapter "Data Visualization" covers the importance and use of visualizing data in order to effectively communicate insights and findings.

- It introduces different types of data visualization, including bar charts, line charts, scatter plots, heat maps, and more, and explains when each is most appropriate to use.
- The chapter also covers how to choose the right visualization for your data, and how to effectively customize and enhance plots to clearly convey information.
- The importance of data storytelling and communication is also discussed, with an emphasis on how to use data visualization to tell a compelling story.
- Real-world examples of data visualization in industry and research are provided to illustrate the practical applications and impact of effective data visualization.
- Additionally, the chapter includes best practices for data visualization, such as keeping in mind audience and context, and avoiding common pitfalls like misleading data representation.
- The importance and impact of data visualization in understanding and communicating complex data sets
- The different types of data visualization and when to use them, including bar charts, line charts, scatter plots, heat maps, and more
- The importance of choosing the right visualization for your data, and how to effectively communicate insights through customization and enhancement of plots
- The role of data storytelling in effectively communicating data insights to a wide audience
- Real-world examples of data visualization in industry and research, including business intelligence, healthcare, and scientific research.
- Tips and best practices for creating effective data visualizations, including design principles and the use of appropriate tools and libraries.
- The potential challenges and limitations of data visualization, and how to overcome them.
- The role of data visualization in decision-making, and the importance of considering the context and audience when creating visualizations.

10.9 TEST YOUR KNOWLEDGE

I. What is the main purpose of data visualization?

a) To make data more presentable
b) To make data more complex
c) To make data more understandable

d) To make data more difficult to understand

II. **What is an example of a univariate visualization?**

a) A scatter plot
b) A bar chart
c) A line chart
d) A heat map

III. **What is an example of a bivariate visualization?**

a) A scatter plot
b) A bar chart
c) A line chart
d) A heat map

IV. **What is an example of a multivariate visualization?**

a) A scatter plot matrix
b) A bar chart
c) A line chart
d) A heat map

V. **What is an example of a non-verbal visualization?**

a) A bar chart
b) A line chart
c) A word cloud
d) A heat map

VI. **What is an example of a sequential color palette?**

a) Blues
b) Reds
c) Greens
d) Purples

VII. **What is an example of a diverging color palette?**

a) Blues
b) Reds
c) Greens
d) Purples

VIII. **What is an example of a qualitative color palette?**

a) Blues
b) Reds

c) Greens
d) Purples

IX. What is an example of a data storytelling technique?

a) Using a scatter plot to show relationships between variables
b) Using a bar chart to show the distribution of a categorical variable
c) Using a narrative to guide the audience through the data and insights
d) Using a line chart to show trends over time

X. What is an example of a customizing technique to enhance plots?

a) Using a scatter plot to show relationships between variables
b) Using a bar chart to show the distribution of a categorical variable
c) Using annotation and labels to provide context
d) Using a line chart to show trends over time

10.10 ANSWERS

I. Answer: c. To make data more understandable
II. Answer: b. A bar chart
III. Answer: a. A scatter plot
IV. Answer: a. A scatter plot matrix
V. Answer: c. A word cloud
VI. Answer: a. Blues
VII. Answer: d. Purples
VIII. Answer: b. Reds
IX. Answer: c. Using a narrative to guide the audience through the data and insights
X. Answer: c. Using annotation and labels to provide context

11

11 DATA ANALYSIS IN BUSINESS

Data analysis in business is the process of using data to make informed decisions and optimize business operations. With the increasing amount of data available to organizations, it has become essential for businesses to develop a strong data analysis strategy. Data analysis allows businesses to gain insights into customer behavior, market trends, and operational efficiencies, which can lead to increased revenue, reduced costs, and improved decision-making. This chapter will explore the various techniques and tools used in data analysis for businesses, including data visualization, statistical analysis, and machine learning. It will also cover the applications of data analysis in different industries and the importance of data governance and ethics in business.

11.1 DATA GOVERNANCE

Data Governance is the overall management of the availability, usability, integrity and security of the data used in an organization. It includes the processes, policies, standards and metrics that ensure the effective and efficient use of data.

Data Governance is critical for any organization that relies on data for decision making, as it helps to ensure that the data is accurate, complete, and reliable. It also helps to ensure that data is used ethically and compliant with laws and regulations.

The main components of Data Governance include:

- Data Governance Framework: This outlines the overall structure and governance of data within the organization, including roles and responsibilities, decision-making processes and compliance requirements.
- Data Governance Policies and Procedures: These provide guidelines for how data should be handled, including data quality, security, privacy and retention.

- Data Governance Metrics: These measure the effectiveness of the data governance program and can be used to identify areas for improvement.
- Data Governance Tools: These are used to support the data governance process, such as data quality tools, data catalogs and data dictionaries.

Data Governance is a continuous process and requires regular review and updates to ensure that it remains effective in meeting the needs of the organization. It is also important to have a designated Data Governance team or individual responsible for overseeing the implementation and maintenance of the Data Governance program.

In business, data Governance is essential for ensuring that data is used effectively to support decision making, compliance and strategic planning. It also helps to ensure that data is protected and used ethically, which is increasingly important in today's data-driven environment.

11.2 DATA QUALITY

Data quality is an important aspect of data analysis in business. It refers to the overall level of accuracy, completeness, consistency, and timeliness of data. In order for data to be considered high quality, it must be accurate and consistent, meaning that it is free of errors and inconsistencies. It must also be complete, meaning that it contains all necessary information, and timely, meaning that it is up-to-date.

Data quality is important in business because it ensures that the data being used for decision making is reliable and accurate. Poor data quality can lead to incorrect conclusions and flawed business decisions. This can result in financial losses and damage to a company's reputation.

To maintain data quality, businesses must have a data governance program in place. This program should include guidelines and procedures for data management, data entry and validation, data cleaning, and data security. Additionally, regular audits should be conducted to ensure that data quality is being maintained.

Data quality can also be improved by implementing data cleaning techniques such as removing duplicate data, correcting errors, and standardizing data formats. Data validation rules can also be put in place to ensure that data is entered correctly and meets certain quality criteria.

11.3 BUSINESS INTELLIGENCE & REPORTING

Business intelligence (BI) and reporting are essential for making data-driven decisions in any business. BI is the process of collecting, storing, and analyzing data to gain insights and make better decisions. Reporting is the process of creating and presenting information in a clear and concise format. Together, BI and reporting allow businesses to turn raw data into actionable insights that can drive growth, reduce costs, and improve operations.

There are several key components to BI and reporting, including data warehousing, data mining, and reporting tools. Data warehousing refers to the process of collecting, storing, and managing large amounts of data, usually from multiple sources, in a central location. This allows for easy access and analysis of the data. Data mining is the process of analyzing large amounts of data to uncover patterns and trends. This can be done using various techniques such as statistical analysis, machine learning, and natural language processing.

Reporting tools are used to create and present the information in a clear and concise format. These tools can range from simple spreadsheets to advanced visualization and dashboard software. They can be used to create a variety of different types of reports, such as financial reports, sales reports, and operational reports.

One of the key benefits of BI and reporting is that it allows businesses to gain a better understanding of their customers, operations, and market trends. This can help to identify areas for improvement and make more informed decisions. Additionally, it can also help to improve collaboration and communication within the organization by providing a common understanding of the data.

However, it is important to note that BI and reporting are only as good as the data that is being analyzed. Therefore, it is crucial to have robust data governance and data quality processes in place to ensure that the data being used is accurate and reliable.

In summary, Business Intelligence (BI) and Reporting are essential for data-driven decision making in any business. It involves collecting, storing, and analyzing data to gain insights and make better decisions. Data warehousing, data mining, and reporting tools are key components of BI. It enables businesses to gain a better understanding of their customers, operations and market trends. However, it's important to have robust data governance and data quality processes in place to ensure accurate and reliable data is being used.

11.4 APPLICATIONS OF DATA ANALYSIS

Data analysis plays a crucial role in the decision-making process of any business. By analyzing large sets of data, businesses can gain valuable insights that can help them improve their operations, increase revenue and stay competitive in the market. In this section, we will discuss some of the key applications of data analysis in the business world.

Sales and Marketing

Data analysis can be used to identify patterns in customer behavior and purchasing habits. This information can be used to target marketing efforts more effectively, increase sales, and improve customer retention. Businesses can use data analysis to segment their customer base, identify key demographics, and create personalized marketing campaigns.

Financial Analysis

Data analysis can be used to identify patterns and trends in financial data. This information can be used to make more informed decisions about investments, budgeting, and forecasting. Businesses can use data analysis to create financial models and simulations, which can help them identify potential risks and opportunities.

Operations and Logistics

Data analysis can be used to optimize the efficiency and effectiveness of business operations. By analyzing data on production processes, inventory management, and supply chain logistics, businesses can identify bottlenecks, reduce costs, and improve delivery times.

Human Resources

Data analysis can be used to gain insight into employee behavior, performance, and engagement. This information can be used to improve recruitment and retention strategies, as well as to identify areas of improvement within the organization.

Risk Management

Data analysis can be used to identify and mitigate potential risks within a business. By analyzing data on financial performance, market trends, and regulatory compliance, businesses can identify potential threats and develop risk management strategies accordingly.

Predictive Analysis

Data analysis can be used to make predictions about future trends and events. This can be useful in a variety of industries, such as finance, healthcare, and retail, to name a few. Predictive analysis can be used to identify potential future market trends and make strategic decisions accordingly.

In conclusion, data analysis is an essential tool for any business. It provides valuable insights that can help businesses to improve their operations, increase revenue, and stay competitive in the market. By understanding the different applications of data analysis in the business world, organizations can make more informed decisions and achieve greater success.

11.5 SUMMARY

- Data analysis in business is a process of using data to gain insights and make informed decisions in the business world.
- Data governance is the process of managing and maintaining the data within an organization. This includes ensuring that data is accurate, consistent, and accessible to those who need it.
- Data quality is the process of ensuring that data is accurate, complete, and reliable. This is important for making accurate and informed decisions based on the data.
- Business intelligence and reporting is the process of using data to inform decision making. This includes creating reports, dashboards, and other visualizations to help stakeholders understand key metrics and trends.
- Applications of data analysis in business include areas such as market research, customer segmentation, and forecasting. This can be used to inform strategy and improve the overall performance of the business.

- Best practices for data analysis in business include creating a strong data governance framework, implementing data quality measures, and regularly reviewing and updating data analysis processes.

11.6 TEST YOUR KNOWLEDGE

I. **What is the main goal of data governance in a business setting?**

a) To ensure data accuracy and integrity
b) To increase sales and revenue
c) To improve customer satisfaction
d) To reduce costs

II. **Which of the following is NOT a component of data quality?**

a) Completeness
b) Timeliness
c) Relevance
d) Profitability

III. **What is the primary purpose of business intelligence and reporting?**

a) To identify trends and patterns in data
b) To generate reports for upper management
c) To improve customer satisfaction
d) To reduce costs

IV. **In what industries is data analysis commonly used?**

a) Finance and banking
b) Retail and e-commerce
c) Healthcare and pharmaceuticals
d) All of the above

V. **What is the main benefit of implementing data governance in a business?**

a) Increased sales and revenue
b) Improved customer satisfaction
c) Better decision making
d) Reduced costs

VI. **What is the main difference between data governance and data management?**

a) Data governance focuses on the overall strategy and policies, while data management focuses on the day-to-day tasks.
b) Data governance is only used in large companies, while data management is used in all companies.
c) Data governance is only used in certain industries, while data management is used in all industries.
d) Data governance is only used for internal purposes, while data management is used for external purposes.

VII. **What is the main goal of data quality?**
a) To increase sales and revenue
b) To improve customer satisfaction
c) To ensure data accuracy and integrity
d) To reduce costs

VIII. **What is the primary role of business intelligence in a company?**
a) To generate reports for upper management
b) To identify trends and patterns in data
c) To improve customer satisfaction
d) To reduce costs

IX. **Which of the following is NOT a common application of data analysis in business?**
a) Market research
b) Fraud detection
c) Inventory management
d) Time travel

X. **Why is data analysis important in business?**
a) It allows companies to make better decisions
b) It helps companies identify trends and patterns in data
c) It improves customer satisfaction
d) All of the above

11.7 ANSWERS

I. Answer: a) To ensure data accuracy and integrity
II. Answer: d) Profitability

III. Answer: b) To generate reports for upper management

IV. Answer: d) All of the above

V. Answer: c) Better decision making

VI. Answer: a) Data governance focuses on the overall strategy and policies, while data management focuses on the day-to-day tasks.

VII. Answer: c) To ensure data accuracy and integrity

VIII. Answer: a) To generate reports for upper management

IX. Answer: d) Time travel

X. Answer: d) All of the above

Appendix A

A. ADDITIONAL RESOURCES FOR FURTHER LEARNING

The chapter "Additional Resources for Further Learning" is designed to provide readers with a comprehensive list of resources that they can use to expand their knowledge and skills in data analysis. This includes books, online tutorials, courses, and more. The goal of this chapter is to provide readers with a wide range of options that they can use to continue their learning journey, regardless of their current level of expertise or experience. Whether you are a beginner looking to learn the basics of data analysis or an experienced professional looking to expand your knowledge and skills, this chapter has something for you.

BOOKS AND EBOOKS

Books and eBooks are a great resource for further learning in data analysis. They offer a comprehensive and in-depth look at various topics, making them a valuable tool for both beginners and experienced practitioners. Here are some popular books and ebooks on data analysis that are worth checking out:

- "**Python Machine Learning**" by Rajender Kumar: This book is a beginner-friendly introduction to machine learning using Python. It covers the basics of data preparation and machine learning and provides hands-on examples for working with data. Available at: https://isbnsearch.org/isbn/9781960833013
- "**R for Data Science**" by Hadley Wickham and Garrett Grolemund: This book is a comprehensive introduction to data science using the R programming language. It covers the full data science process, including data manipulation, visualization, and modeling.
- "**Data Smart: Using Data Science to Transform Information into Insight**" by John W. Foreman: This book provides a practical introduction to data science for

business professionals. It covers the basics of data analysis and provides examples of how it can be applied to real-world business problems.

- "**Data Wrangling with Python**" by Jacqueline Kazil and Katharine Jarmul: This book is a hands-on guide to data wrangling, the process of cleaning, transforming, and analyzing data. It covers the basics of Python and the popular pandas library and provides examples of how to work with different types of data.

These are just a few examples of the many books and ebooks available on data analysis. As you continue to learn and explore the field, be sure to check out additional resources to help you expand your knowledge and skills.

WEBSITES AND BLOGS

Websites and Blogs are a great resource for further learning on data analysis. Some popular websites and blogs that offer valuable information and tutorials on data analysis include:

- **Kaggle** (https://www.kaggle.com/) - A platform for data science competitions, with a wide range of datasets and resources for learning data analysis.
- **DataCamp** (https://www.datacamp.com/) - An interactive learning platform for data science and data analysis, offering tutorials and courses in R and Python.
- **Dataquest** (https://www.dataquest.io/) - A data science and data analysis learning platform, with a focus on hands-on projects and real-world examples.
- **Data Science Central** (https://www.datasciencecentral.com/) - A community-driven website with a wide range of articles, tutorials, and resources for data science and data analysis.
- **KDnuggets** (https://www.kdnuggets.com/) - A website that offers a wide range of resources, including tutorials, articles, and news on data science and data analysis.
- **Data Science Society** (https://datascience.community/) - A community-driven website, where data scientists share their knowledge, experience and collaborate on data science projects.

These websites and blogs offer a wealth of information on data analysis, including tutorials, code snippets, and real-world examples. They are a great resource for anyone looking to deepen their understanding of data analysis and take their skills to the next level.

COMMUNITY FORUMS AND GROUPS

In addition to books and websites, community forums and groups can also be a great resource for further learning about data analysis. These forums and groups provide a platform for individuals to ask questions, share knowledge and experiences, and collaborate on projects. Some popular options include:

- Reddit's Data Science community (https://www.reddit.com/r/datascience/) - This subreddit is a great place to ask questions, share resources, and connect with other data scientists.
- Data Science Society (https://datascience.community/) - This is a community-driven platform where data scientists can ask questions, share knowledge and collaborate on projects.
- Data Science Group (https://www.facebook.com/groups/datasciencegroup/) - This is a Facebook group for data scientists, analysts, and engineers to share knowledge, ask questions, and collaborate on projects.

By participating in these community forums and groups, you can stay up-to-date on the latest trends and techniques in data analysis, as well as connect with other professionals in the field.

ONLINE COURSES AND CERTIFICATIONS

Online courses and certifications can be a great way to learn about data analysis in depth. Some popular platforms for online learning include Coursera, Udemy, and edX. These platforms offer a wide range of courses, from beginner-friendly introductions to more advanced topics.

One popular course for learning data analysis is the **Data Science Specialization** offered by Johns Hopkins University on Coursera. This specialization includes nine courses that cover the basics of data science, including data visualization, probability and statistics, and machine learning.

Another great resource is DataCamp, which offers interactive data science and analytics courses. Their courses are designed to help you learn by doing, and they cover a wide range of topics, including Python, R, and SQL.

Datacamp is available for free with a 7-day trial, after that it is a paid service.

For certification, Data Science Council of America (DASCA) offers a number of certification programs including data analyst, data scientist, and big data engineer.

In addition, IBM offers a Data Science Professional Certification program, which includes a series of online courses and a final exam. This certification is aimed at professionals who want to demonstrate their skills in data science, machine learning, and Python programming.

The links to the above resources are:

- Johns Hopkins University's Data Science Specialization on Coursera: https://www.coursera.org/specializations/jhu-data-science
- DataCamp: https://www.datacamp.com/
- Data Science Council of America (DASCA): https://dasca.org/
- IBM Data Science Professional Certification: https://www.ibm.com/cloud/garage/content/course/data-science-professional-certificate/

DATA ANALYSIS CONFERENCES AND MEETUPS

Data Analysis Conferences and Meetups are events that bring together professionals, researchers, and academics in the field of data analysis to share their knowledge and experience, as well as learn from others. These events provide an excellent opportunity to network with other professionals in the field, hear about the latest research and developments, and learn about new tools and techniques. Some popular data analysis conferences and meetups include:

- Strata Data Conference: Strata Data Conference is an annual event that brings together data professionals from around the world to share their knowledge and experience. The conference features keynotes, tutorials, and sessions on a wide range of topics, including data science, big data, data engineering, and more. https://conferences.oreilly.com/strata/strata-ny
- Data Science Summit: The Data Science Summit is an annual event that focuses on the latest trends and developments in data science, machine learning, and AI. The summit features keynotes, tutorials, and sessions on a wide range of topics, as well as an exhibition hall where attendees can learn about the latest data science tools and technologies. https://www.datasciencesummit.com/

- Data Science Conference: The Data Science Conference is an annual event that focuses on the latest trends and developments in data science, machine learning, and AI. The conference features keynotes, tutorials, and sessions on a wide range of topics, as well as an exhibition hall where attendees can learn about the latest data science tools and technologies. https://www.datascienceconf.com/
- Data Science Meetup: Data Science Meetup is a community-driven event that brings together data scientists, engineers, analysts, and researchers to share their knowledge and experience. The meetup features presentations, discussions, and networking opportunities, as well as an opportunity to learn about the latest tools and techniques in data science. https://www.meetup.com/topics/data-science/
- Big Data and AI Conference: The Big Data and AI Conference is an annual event that brings together professionals from the big data and AI communities to share their knowledge and experience. The conference features keynotes, tutorials, and sessions on a wide range of topics, including big data, AI, machine learning, and more. https://www.bigdata-ai.com/

These are just a few examples of the many data analysis conferences and meetups that take place around the world. By attending these events, you can learn from experts in the field, network with other professionals, and stay up-to-date on the latest trends and developments in data analysis.

DATA ANALYSIS TOOLS AND SOFTWARE

Data analysis tools and software are essential for effectively processing, analyzing, and interpreting large sets of data. There are a wide variety of tools available, each with their own unique features and capabilities. Here is a list of some popular data analysis tools and software, along with a brief explanation of their features:

- **R**: R is an open-source programming language and software environment for statistical computing and graphics. It is widely used for data analysis, visualization, and modeling. R has a large community of users and developers, and there are many packages available for a wide range of data analysis tasks.
- **Python**: Python is another open-source programming language that is widely used for data analysis. It has a large ecosystem of libraries and frameworks, such as NumPy,

pandas, and scikit-learn, that provide powerful tools for data manipulation and analysis.

- **SAS**: SAS is a proprietary software suite for advanced analytics, business intelligence, and data management. It includes a wide range of tools for data analysis, visualization, and reporting. SAS is widely used in business, finance, and government organizations.
- **SPSS**: SPSS is a statistical analysis software that provides a wide range of tools for data analysis, including descriptive statistics, inferential statistics, and data visualization. It is widely used in social science research and in the business industry.
- **Excel**: Microsoft Excel is a widely used spreadsheet software that has built-in data analysis tools, such as pivot tables, charts, and graphs. It is commonly used for data cleaning, data manipulation, and basic data analysis tasks.
- **Tableau**: Tableau is a data visualization tool that allows users to easily create interactive dashboards and reports. It provides a drag-and-drop interface for data exploration and analysis, making it a popular choice for business intelligence and data discovery.
- **RapidMiner**: RapidMiner is an open-source data science platform that provides a wide range of tools for data preparation, machine learning, and model evaluation. It is widely used in business and industry for predictive analytics and data mining.
- **KNIME**: KNIME is a free and open-source data analytics platform that provides a wide range of tools for data manipulation, visualization, and modeling. It has a user-friendly interface and is widely used in business and industry for data integration, machine learning, and predictive analytics.
- **Alteryx**: Alteryx is a data analytics platform that provides a wide range of tools for data preparation, visualization, and modeling. It is widely used in business and industry for data integration, machine learning, and predictive analytics.
- **Power BI**: Power BI is a business intelligence tool developed by Microsoft. It allows users to connect to various data sources, create interactive visualizations and reports, and share them with others.

CNOCLUSION

In conclusion, the field of data analysis is constantly evolving and it is crucial for professionals and enthusiasts to continue their education and stay up-to-date with the latest developments.

Whether you are new to the field or a seasoned professional, there are a plethora of resources available to help you further your knowledge and skills.

Books and ebooks, websites and blogs, community forums and groups, online courses and certifications, data analysis tools and software, conferences and meetups, and case studies and examples are all valuable resources for learning and staying current in the field. Additionally, exploring career opportunities and professional organizations can also be beneficial for those looking to advance their career in data analysis.

Learning and staying current in the field of data analysis is an ongoing process and should be a priority for those looking to excel in their careers. By utilizing the resources provided in this chapter, you can take the first step on your journey to becoming a well-rounded and proficient data analyst.

Appendix B

B. INSIDER SECRETS FOR SUCCESS AS A DATA ANALYST

In this chapter, we will discuss some valuable tips and best practices that can help you succeed in the field of data analysis. Whether you are a beginner or an experienced professional, these tips will provide valuable guidance on how to approach data analysis projects, overcome common challenges, and achieve success in your work. From understanding the importance of clear communication and visualization, to leveraging the power of automation and collaboration, these tips will provide valuable insights and strategies for achieving success in the field of data analysis.

TIPS FOR SUCCESS IN DATA ANALYSIS

The field of data analysis is constantly evolving, and it is essential to stay up-to-date with the latest developments and techniques in order to be successful. In this section, we will discuss some of the key tips and strategies that can help you succeed in data analysis.

- **Develop a strong foundation in statistics and mathematics**: A strong understanding of statistics and mathematics is essential for success in data analysis. This includes understanding concepts such as probability, distributions, and statistical inference.
- **Learn the most commonly used data analysis tools**: There are a variety of data analysis tools available, and it is important to become proficient in the most widely used ones such as Excel, R, and Python.
- **Learn to work with big data**: The amount of data available to organizations is growing at an unprecedented rate, and learning to work with big data is becoming increasingly important for data analysts.

- **Learn to communicate effectively**: Data analysis is not just about crunching numbers, it's also about communicating your findings and insights to others. Effective communication is critical for success in this field.
- **Practice data visualization**: The ability to create clear and effective visualizations is essential for data analysis. The ability to present data in a way that is easy to understand is a key skill that can help you succeed in this field.
- **Stay up to date with the latest developments**: The field of data analysis is constantly evolving, and it is important to stay up-to-date with the latest developments in order to stay ahead of the curve.
- **Practice, practice, practice**: The more you practice data analysis, the better you will become. Look for opportunities to practice your skills, whether it's through internships, projects, or personal projects.

By following these tips, you can increase your chances of success in data analysis and become a valuable asset to any organization. Remember, the key to success in this field is to always be learning and growing, and to never stop challenging yourself.

DATA ANALYSIS CAREERS AND PROFESSIONAL RESOURCES

Data analysis is a rapidly growing field with a wide range of career opportunities. From data scientists and business analysts to data engineers and data architects, there are many different roles that involve working with data to extract insights and inform business decisions.

One of the key skills needed for a career in data analysis is the ability to work with various data analysis tools and software. Some of the most popular tools include Python, R, SQL, SAS, and Tableau. Having experience with these tools and being proficient in at least one programming language is essential for many data analysis roles.

Another important skill is the ability to communicate effectively with non-technical stakeholders, such as business leaders and managers. Being able to present data insights in a clear and actionable way is crucial for making data-driven decisions.

In addition to these technical skills, data analysts should also have a strong understanding of statistics and data modeling, as well as domain-specific knowledge in the industry they are working in.

There are a number of professional organizations and associations that support data professionals in their careers, such as The Data Warehousing Institute (TDWI), the International Institute for Analytics (IIA), and the Data Science Society (DSS). These organizations provide resources, networking opportunities, and training and certification programs for data professionals.

The job market for data analysts is expected to continue growing in the coming years, with many companies looking to hire professionals to help them make sense of their data and gain a competitive edge. Some of the top industries hiring data analysts include technology, finance, healthcare, and retail.

Overall, a career in data analysis can be highly rewarding, with the opportunity to work on challenging and meaningful projects, and to make a real impact on the success of a business.

FIND A JOB AS A DATA ANALYST

Finding a job as a data analyst can be a bit of a challenge, but with the right approach, it's definitely possible. Here are a few tips to help you land your dream job in data analysis:

- **Build a strong resume and portfolio**: Your resume and portfolio are your first impressions to potential employers, so make sure they are polished and professional. Highlight your relevant skills, experience, and education, and include examples of your work to showcase your abilities.
- **Network**: Networking is key when it comes to finding a job. Attend industry events, join professional organizations, and connect with people in the field. You never know who might have a job opening or know someone who does.
- **Learn the right skills**: Data analysis is a field that's constantly evolving, so it's important to stay up-to-date with the latest tools and techniques. Take online courses, read industry publications, and invest in yourself to learn the skills that are most in-demand.
- **Be flexible**: Be open to different types of data analysis jobs. You may have to start in a junior or entry-level position, but this can be a great way to gain experience and work your way up to more senior roles.

- Show your passion: Employers want to see that you're passionate about data analysis. Show this through your work, your blog posts or publications and your willingness to learn and continuously improve.
- **Use the right keywords in your resume and cover letter**: Be sure to use relevant keywords and terms that employers are looking for in your resume and cover letter. This will help your application get noticed by the right people.
- **Make sure your LinkedIn profile is up-to-date**: Your LinkedIn profile is a great way to showcase your professional experience and connect with potential employers. Be sure to keep it up-to-date, and consider creating a portfolio on LinkedIn to showcase your work.
- **Keep an open mind**: Be open to different types of companies and industries, as data analysis can be applied in a variety of different fields.
- **Be patient**: Finding the perfect job may take some time, so be patient and don't get discouraged. Keep applying and networking, and eventually you'll find the right opportunity.
- **Prepare for the interview**: Research the company and the position, and practice your interviewing skills. Be prepared to discuss your qualifications, skills, and experience in detail, and be ready to answer questions about your approach to data analysis.

By following these tips, you'll be well on your way to finding a job as a data analyst. Remember to stay positive, stay persistent, and keep learning, and you'll eventually find the right opportunity.

Appendix C

C. GLOSSARY

Here are some of the key terms and definitions included in the glossary:

A

- **Aggregate**: A summary or total of a set of data.
- **Algorithm**: A set of instructions for solving a problem or performing a task.
- **Analysis**: The process of studying and interpreting data to extract useful information.
- **API**: Application Programming Interface, a set of tools and protocols for building software and applications.

B

- **Big Data**: A term used to describe large, complex sets of data that cannot be easily analyzed using traditional methods.
- **Business Intelligence (BI)**: The use of technologies, applications, and practices for the collection, integration, analysis, and presentation of business information.
- **Box and Whisker Plot**: A graphical representation of a set of data that shows the minimum, first quartile, median, third quartile, and maximum values.

C

- **Clustering**: A method of grouping similar data points together based on their characteristics or attributes.
- **Correlation**: A measure of the relationship between two or more variables.
- **Central Tendency**: A measure of the central or typical value in a dataset. Mean, median, and mode are examples of measures of central tendency.

D

- **Data Storytelling**: The process of using data visualization and other techniques to communicate insights and information to an audience.
- **Dimensionality Reduction**: A technique used to reduce the number of variables or features in a dataset.
- **Data Governance**: The overall management and oversight of the availability, usability, integrity, and security of an organization's data.
- **Data Quality**: The degree to which a set of characteristics of data meets the requirements of its intended use.
- **Data Analysis**: The process of inspecting, cleaning, transforming, and modeling data with the goal of discovering useful information, suggesting conclusions, and supporting decision-making.
- **Data Visualization**: The use of graphical representations to make data more understandable and accessible.
- **Dimensionality Reduction**: The process of reducing the number of variables in a dataset while retaining as much information as possible.
- **Distribution**: A description of the way in which a variable is spread out across its possible values.
- **Descriptive Statistics**: The branch of statistics that deals with summarizing and describing data.

E

- **Exploration**: The process of examining and analyzing data to discover patterns, trends, and insights.
- **Exploratory Data Analysis (EDA)**: The initial examination of a dataset to identify patterns, trends, and outliers.

F

- **Frequency Distribution**: A table or graph that shows the number of occurrences of each value in a dataset.

H

- **Hypothesis Testing**: A statistical method used to test the validity of a claim or hypothesis about a population.
- **Histogram**: A graphical representation of a frequency distribution in which the horizontal axis represents the values of a variable and the vertical axis represents the frequency of observations.
- **Histogram**: A graph that shows the distribution of a variable by dividing the range of possible values into bins and counting the number of observations that fall into each bin.

I

- **Inferential Statistics**: A branch of statistics that deals with making predictions or inferences about a population based on a sample of data.

k

- **Kurtosis**: A measure of the peakedness or flatness of a probability distribution compared to the normal distribution.

M

- **Machine Learning**: A type of artificial intelligence that allows a system to learn and improve from experience.
- **Mean**: A measure of central tendency that is calculated by summing all the values in a dataset and dividing by the number of values.
- **Median**: A measure of central tendency that is the middle value in a dataset when the values are arranged in order.
- **Mode**: A measure of central tendency that is the value that occurs most frequently in a dataset.
- **Measures of Association**: Statistical techniques that are used to measure the strength and direction of the relationship between two variables.
- **Measures of Correlation**: Statistical techniques that are used to measure the strength and direction of the relationship between two variables.
- **Measures of Spread**: A set of statistics that describe the degree of variation or dispersion in a set of data.

N

- **Normalization**: A technique used to scale or adjust the values in a dataset to a common range or scale.

O

- **Outlier**: A value that is significantly different from the rest of the values in a dataset.

P

- **Predictive Modeling**: The process of using statistical or machine learning techniques to build a model that can predict future outcomes based on historical data.
- **Prescriptive Analytics**: A type of advanced analytics that uses data, models, and algorithms to provide advice and recommendations for decision-making.
- **Probability**: The likelihood or chance of a particular event occurring.
- **Python**: A high-level programming language used for a wide range of tasks such as web development, scientific computing, data analysis, and artificial intelligence.
- **Predictive Analytics**: The use of data, statistical algorithms, and machine learning techniques to identify the likelihood of future outcomes based on historical data.

R

- **Regression**: A statistical method used to model the relationship between a dependent variable and one or more independent variables.

S

- **Sample**: A subset of a population used to represent the population in statistical analysis.
- **Scaling**: A technique used to adjust the values in a dataset to a common range or scale.
- **Standard Deviation**: A measure of spread that describes the amount of variation or dispersion in a dataset.
- **Statistics**: The branch of mathematics that deals with the collection, analysis, interpretation, presentation, and organization of data.
- **Skewness**: A measure of the asymmetry of a probability distribution around its mean.

- **Standard Deviation**: A measure of the spread of a set of values, calculated as the square root of the variance.

V

- **Visualization**: The process of using graphs, charts, and other visual aids to represent and communicate data.
- **Variance**: A measure of spread that describes the amount of variation or dispersion in a dataset.
- **Variable**: A characteristic or attribute of a dataset that can take on different values.

X

- **X-axis**: The horizontal axis of a graph or plot.

Y

- **Y-axis**: The vertical axis of a graph or plot.

This glossary is just a small sample of the terms and definitions included in the book. By familiarizing yourself with these terms and others in the glossary, you will have a better understanding of the concepts and techniques covered in the book.

A HUMBLE REQUEST FOR FEEDBACK!

Dear Readers,

I hope this message finds you well. I am reaching out to kindly request your feedback on my recent book titled "**Mastering Data Analysis with Python**" which is a comprehensive guide for individuals looking to master data analysis techniques using Python.

As an author, I strive to create high-quality content that is both informative and easy to understand. I believe that this book is an excellent resource for anyone looking to improve their data analysis skills, whether they are a beginner or an experienced professional. It covers a wide range of topics, including NumPy, Pandas, and Matplotlib, which are some of the most popular libraries used in Python for data analysis.

I would greatly appreciate it if you could take the time to leave a review of my book. Your feedback will help me to improve my writing and create even better content for my readers in the future. Thank you for your time, and I hope that you found my book to be informative and useful.

Best regards,

Rajender Kuma

Made in the USA
Coppell, TX
25 June 2023

18503887R00188